Black Holes and Warped Spacetime

Black Holes and Warped Spacetime

William J. Kaufmann, III

Department of Physics
San Diego State University

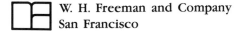 W. H. Freeman and Company
San Francisco

Sponsoring Editor: Arthur C. Bartlett; *Project Editor:* Pearl C. Vapnek; *Copyeditor:* Karen Judd; *Designer:* Marjorie Spiegelman; *Production Coordinator:* Linda Jupiter; *Illustration Coordinator:* Cheryl Nufer; *Artist:* Dale Johnson; *Compositor:* Graphic Typesetting Service; *Printer and Binder:* The Maple-Vail Book Manufacturing Group.

Library of Congress Cataloging in Publication Data

Kaufmann, William J
    Black holes and warped spacetime.

    Includes index.
    1. Black holes. 2. Space and time. I. Title.
QB843.B55K38   520   79–18059
ISBN 0–7167–1152–4
ISBN 0–7167–1153–2 pbk.

Printed in the United States of America

9 8 7 6 5 4

Cover art: An artist's conception of the binary star system Cygnus X-1. Painting by Michael Cedeno.

once again for Lee
with all my love

Contents

Preface

A black hole is one of the most fantastic things ever predicted by modern science. It is a place where gravity is so strong that nothing — not even light — can escape. It is a place where gravity is so strong that a hole has been rent in the very fabric of space and time. Surrounding this yawning chasm is a "horizon" in the geometry of space where time itself stands still. And inside this hole, beyond this horizon, the directions of space and time are interchanged.

There are wormholes to other universes, time tunnels and time machines that would bring you back to where you started before you left. There is antigravity beyond the hole's center. And around the hole, there are regions where you have to travel at the speed of light in order to remain in the same place.

As incredible as it may seem, we are actually finding these things in the sky. We are finding them as the corpses of massive stars. And we are also finding them at the centers of exploding galaxies: enormous black holes containing billions of completely collapsed suns.

Strangest of all, our understanding of the birth of the universe leads us to believe that vast numbers of very tiny black holes should have been created during the first few milliseconds after the Big Bang. Unlike their massive and supermassive cousins, these tiny black holes evaporate and explode by randomly vomiting particles and radiation.

During the 1920s, scientists finally figured out how energy is produced at the sun's center. They could finally answer the childlike question, "Why does the sun shine?" in terms of thermonuclear reactions. At a news conference describing these nuclear processes, a famous scientist was asked if humanity might ever unleash or tap the energy of the nucleus here on the earth. "Obviously, never!" was the response. Such possibilities were relegated to the realm of science fiction. But not for long.

In an advanced technological society, it is crucial that the citizenry be aware of advances at the frontiers of science. This is especially true in a democracy, where the people are the electorate. Although many topics in black hole physics are still in the realm of science fiction, they may not remain there for long.

Knowledge of the universe bestows awesome power. To understand the secrets of atoms and galaxies is to become like the gods. And we fly to the moon, light the fires of the stars, and perhaps someday probe a black hole. Whether we use these abilities for the betterment of humanity or for the devastation of our planet is entirely a matter of our own free choice. The profound laws of nature are not evil. Only our intentions and motivations are sometimes malevolent. This is why it is so important to have an informed public. Only with an enlightened electorate do we stand a chance of making intelligent decisions.

August 1979   William J. Kaufmann, III

Black Holes and
Warped Spacetime

1

The Evolution of Stars

Somewhere in our Galaxy, in a far-off place many thousands of light years away, a vast cloud of interstellar gas and dust drifts silently across the near-perfect vacuum. The wispy edges of the cloud stretch outward into the blackness for trillions of miles in all directions, as if to reach for the stars beyond.

Like many other clouds scattered around our Galaxy, this interstellar apparition contains an enormous amount of matter. Composed mostly of hydrogen and helium, the cloud contains more than enough mass to create dozens upon dozens of stars like our sun. But no stars shine here. The process of starbirth has yet to occur.

In spite of its enormous mass, the cloud has such vast dimensions that the atoms are spread thin across the cloud's colossal volume. Indeed, if you journeyed even to the very center of this cloud, you might first think that you were still in the near-perfect vacuum of interstellar space. But with careful examination, you would find roughly ten atoms per cubic centimeter (a cubic centimeter is about the size of a cube of sugar). In sharp contrast, the air we breathe here on earth contains 30 billion billion atoms in each cubic centimeter.

The majority of atoms you would find in this diaphanous cloud are hydrogen, by far the most abundant gas in the universe. And for every sixteen hydrogen atoms you would find a helium atom, the second most abundant element. Only with a great deal of searching could you find atoms of the heavier elements, such as carbon, nitrogen, oxygen, and iron, to name a few.

And it is cold! At only 100 degrees above absolute zero (about $-173°C = -280°F$), the atoms of the cloud meander leisurely, hardly ever colliding. This is the essence of a stellar womb, a quiet place in our Galaxy that will soon be ablaze with newborn stars. But for now the cloud waits patiently, as it has for millions of years, for the approach of a spiral arm.

We live in a spiral galaxy, a huge rotating collection of gas, dust, and hundreds of billions of stars. To view our Galaxy, you would have to travel far out into space, far beyond the most remote stars you can see in the sky. And after journeying hundreds of thousands of light years, you could turn and look back to see our celestial whirlpool stretched out before you.

There are countless millions of spiral galaxies scattered across the depths of space. Viewed from afar, our Galaxy would probably look quite similar to the galaxy called M81, shown in Figure 1-1. A typical spiral galaxy like our own measures 100,000 light years in diameter. Our star, the sun, is located two-thirds of the way from the center to the edge, between two arching spiral arms. These spiral arms rotate majestically about the Galaxy's nucleus, carrying shock waves that compress the interstellar material. This compression triggers the birth of the stars.

As a spiral arm sweeps through an interstellar cloud, the widely spaced atoms are suddenly jostled close together. The cloud had been transparent. But now, with the atoms closer together, feeble starlight can no longer penetrate. Our interstellar cloud has become the *dark nebula.*

Dark nebulas are hard to find. They show up best when silhouetted against a background field of stars. An excellent example is seen in Figure 1-2, and a color photograph of the famous Horsehead Nebula is shown in Plate 1.

Since the dark nebula is opaque, light from distant stars no longer penetrates the compressed cloud to warm its gases. The temperature plunges toward absolute zero. As the temperature falls, the atoms move more slowly than ever, so slowly that the weak force of gravity between individual atoms now begins to dominate the nebula's internal structure.

As you may well expect, the dark nebula is not perfectly smooth and uniform. Instead, as luck would have it, there are certain locations where there are a few more atoms than average. And there are other locations where there are fewer atoms than average.

All matter has gravity. The more matter there is at a particular place, the stronger is the gravity around that place. Consequently, at the locations inside the dark nebula where, by chance, there is a slight excess of atoms, there is also a slightly stronger gravitational field. These locations of extra gravity easily attract slowly moving nearby atoms. As the number of atoms begins to grow, the gravity at these locations becomes still stronger, thereby attracting even more material from the surrounding nebula. In this way, the nebula begins to break up into lumps or *globules.*

At its formation, a typical globule may be several billion miles in diameter and contain an amount of matter a few times the mass of the sun (the sun's mass is nearly 2 billion billion billion tons).

Figure 1-1   The Spiral Galaxy M81 (also called NGC 3031)
Our Galaxy, if you could view it from afar, would probably look like this galaxy in the constellation of Ursa Major. A typical spiral galaxy is 100,000 light years in diameter and contains about 200 billion stars. (Kitt Peak National Observatory.)

Figure 1-2   Dark Nebulas in Ophiuchus
Thousands upon thousands of stars are seen in this photograph of a very rich region in the constellation of Ophiuchus. The dark areas are caused by foreground clouds of cool gas and dust. These vast clouds obscure the light from the stars beyond. (Yerkes Observatory.)

Numerous globules are seen silhouetted against brighter background nebulosity in Figure 1-3. Globules also appear in the color photograph of the Eagle Nebula in Plate 2.

A globule is unstable. Quite simply, the globule is incapable of supporting its own weight. Trillions upon trillions of tons of gas pressing inward from all sides cause the globule to contract. Under the relentless influence of its own gravity, the globule becomes smaller and smaller, squeezing the gases at the center of the contracting sphere to even higher pressures and densities.

In most circumstances in nature, pressure and temperature go hand in hand. Consequently, as the pressures at the core of the contracting globule rise, the temperature begins to climb. With increasing temperature, the gases deep inside the shrinking globule begin to glow. Radiation begins to filter outward through the contracting sphere of gas. You soon notice that the globule is no longer dark. The first light to make its way to your eyes is a dull red, and the gases shimmer faintly like glowing embers in a fireplace. The globule has transformed itself into a *protostar.*

But the protostar is also unstable against gravity. This sphere of gas is still unable to support the tremendous weight of its outer layers. The protostar, therefore, continues to contract, pushing the temperatures and pressures of the interior gases higher and higher.

Finally, when the temperature at the protostar's center reaches 10 million degrees, *hydrogen burning* is ignited. At this temperature, the nuclei of the hydrogen atoms are moving so swiftly that when they collide, they permanently stick together. In this remarkable process, hydrogen is actually converted into helium. For every four hydrogen nuclei that are fused together, a helium nucleus is created. But even more importantly, the resulting helium nucleus weighs slightly less than the four hydrogen nuclei from which it was created. Some matter has disappeared. This missing material has been converted into pure energy according to Albert Einstein's famous equation $E = mc^2$. This process, called a *thermonuclear reaction,* is by far one of the most powerful processes to occur in nature. It is the direct transformation of mass into energy. The enormous release of energy that accompanies hydrogen burning therefore creates conditions by which the protostar can finally support the weight of its outer layers. Contraction is halted. A star is born.

Figure 1-3   Globules
*Numerous globules are seen silhouetted against the brighter
background nebulosity. Globules represent the earliest stages in the
birth of stars. In only a few million years, this region of space will
be ablaze with dazzlingly bright newborn stars. (Lick Observatory.)*

A newborn star is a star that has recently ignited hydrogen burning at its center. Searching the heavens, you can find many of these young stars scattered across space. They are easily identified because they are usually still embedded in vast sweeping fragments of the clouds from which they emerged. Often these gases shine with unprecedented beauty as intense ultraviolet light from young massive stars causes the interstellar gases to fluoresce and glow. Superb examples of these stellar nurseries include the famous Orion Nebula shown in Figure 1-4 and the Lagoon Nebula seen in Figure 1-5.

When you glance up at the sky, almost every star you see is a young star still consuming hydrogen fuel at its core. Our star, the sun, is a fine example, having been born a mere 5 billion years ago. Six hundred million tons of hydrogen are transformed into helium *each second* at the sun's center. This may sound like a frightening rate, but the sun is in no danger of running out of hydrogen fuel. The sun's mass is so huge that it has ample fuel to continue hydrogen burning for yet another 5 billion years.

Although we usually speak of hydrogen "burning," nothing is "burned" in the usual sense of a candle flame or logs in a fireplace. Instead, the thermonuclear reactions deep inside stars actually fuse the nuclei of atoms together. In these thermonuclear furnaces, lighter elements are transformed into heavier elements at inconceivable temperatures and pressures.

Stars like the sun burn hydrogen at their cores for billions of years. Hydrogen is the fuel and helium is the ash. But gradually, as these stars move from adolescence into adulthood, the amount of helium at their centers builds up while the supply of hydrogen dwindles. And finally, at a crucial stage in the lives of all stars, the time arrives when the hydrogen is depleted. This triggers major changes in the structure and appearance of a star, changes more profound and dramatic than anything the star has known since its birth eons in the past.

Imagine one of these mature stars, like the sun, 5 billion years from now. As with all stars, it takes nearly a million years for energy created at the star's center to make its way laboriously to the star's edge and to escape finally as starlight. Consequently, although significant changes are about to occur at the star's core, you will not

Figure 1-4 **The Orion Nebula (also called M42 or NGC 1976)**
Several newborn stars are buried deep within this nebula. Intense ultraviolet light from these stars causes the surrounding gases to glow. This nebula is just barely visible to the naked eye as the middle "star" in the "sword" of the constellation of Orion. (Lick Observatory.)

Figure 1-5   The Lagoon Nebula (also called M8 or NGC 6523)
*Globules, protostars, and newborn stars exist side by side in this
beautiful nebula in the constellation of Sagittarius. As with the
Orion Nebula, ultraviolet light from young stars is causing the
gases to fluoresce. While the Orion Nebula is only 1,500 light years
away, the distance to the Lagoon Nebula is thought to be about
6,500 light years. (Kitt Peak National Observatory.)*

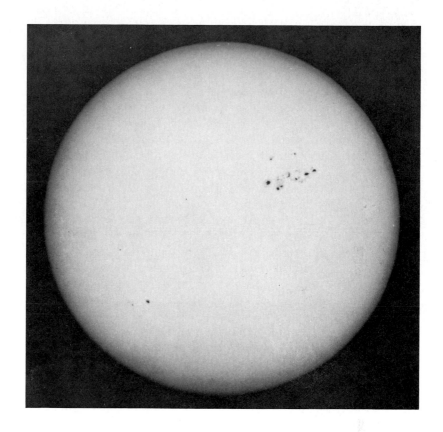

Figure 1-6   *The Sun*
*The sun is a young star, having been formed only 5 billion years
ago. At the sun's center, hydrogen is fused into helium in a
thermonuclear reaction called "hydrogen burning." There is enough
hydrogen fuel at the sun's core to continue hydrogen burning for
another 5 billion years. (Kaufmann Industries, Inc.)*

notice anything unusual for quite a while. There is still a million years' worth of radiation climbing toward the star's surface.

When all the hydrogen at a star's center is used up, hydrogen burning shuts off. With no more outpouring of energy, the star's core becomes unstable against the influence of gravity. Unable to support its own weight, the star's helium-rich core begins to shrink. As this core becomes more and more compressed, temperatures deep inside the star rise to new heights. Although no more hydrogen remains at the star's center, there is still plenty of fresh hydrogen fuel between the core and the surface. Finally, temperatures surrounding the collapsing core rise high enough to ignite hydrogen burning in a thin *shell* around the core.

Observing this stellar metamorphosis from afar, you notice nothing unusual at first. But with the ignition of *shell hydrogen burning,* the star has found a new source of energy and a fresh supply of fuel. Gradually, almost imperceptibly at first, the star begins to swell. The star's outer layers are slowly pushed outward as the star re-adjusts to the continued contraction of its core, now surrounded by a thin shell of hydrogen burning.

The gravitational compression of the helium-rich core forces temperatures and pressure at the star's center to unprecedented heights. Finally, at 100 million degrees, the helium nuclei at the star's center are moving so fast and colliding so violently that they fuse together to form carbon and oxygen. For the first time in the star's history, a new thermonuclear reaction called *helium burning* makes its appearance.

The ignition of helium burning produces a new outpouring of energy that prevents any further core contraction. Now the star has two thermonuclear reactions occurring deep in its interior: helium burning at the center surrounded by a shell of hydrogen burning. In dramatic response to this double-barreled source of energy, the star swells to enormous proportions, increasing its volume a billionfold. As the outer layers of the star are pushed farther and farther outward, the atoms of which these layers are composed get farther and farther apart. The density and pressure in the star's outer layers decrease, and in response, so does the temperature. As you watch from a safe distance, the bloated star's surface has become big and cool.

The surface temperature of a star like the sun is 6,000 degrees, and the white-hot gases shine with a dazzling white light. In the remote future, when the sun expands to an enormous size, its surface temperature will drop to 3,000 degrees. Instead of being white-hot, the cooler surface gases will then glow with a reddish hue, like the coals in a fireplace or a horseshoe in a blacksmith's hearth. Stars of this type are appropriately called *red giants.*

Five billion years from now, as our sun evolves into a red giant, sun-scorched Mercury will be the first planet to be vaporized. As Mercury is swallowed by the approaching solar surface, Venus's thick carbon dioxide atmosphere will be swept away while Earth's oceans boil. Venus and Earth soon suffer Mercury's fate as rocks melt and turn to gas. At the end of the transformation, the red giant sun will have a diameter of 200 million miles, slightly larger than Earth's orbit. Even if our planet could survive, it would be inside the swollen sun's reddish atmosphere.

Almost every reddish star you see in the sky is a red giant. Aldebaran (in Taurus), Antares (in Scorpius), Arcturus (in Boötes), and Betelgeuse (in Orion) are fine examples. All of them have hot, dense cores, burning helium and hydrogen. All of them have enormous, wispy, bloated atmospheres glowing in shades of blood red.

Just as helium is the ash of hydrogen burning, carbon and oxygen are the ashes of helium burning. And as you may expect, after a few billion years as a red giant, all of the helium at the star's center is used up. Helium burning, therefore, shuts off and the star's core once again becomes unstable and again begins to contract still further under the influence of gravity. Temperatures and pressures are driven higher than ever before. Finally, temperatures above the carbon – oxygen-rich core are high enough to ignite helium burning in a thin shell, the same helium that was left behind from the shell of hydrogen burning that has been working its way outward. At this stage, the star has two thin shells of thermonuclear reactions: an inner helium burning shell and an outer hydrogen burning shell, as shown schematically in Figure 1-8.

This is the last stage in the life of a star like the sun. Stars like the sun do not possess enough mass to ignite any further thermonuclear reactions. The weight of the star's outer layers is not sufficient to

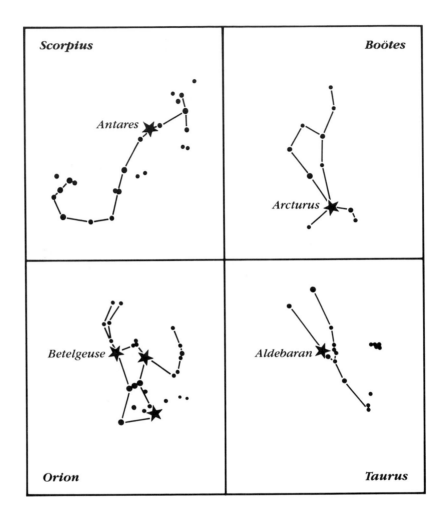

Figure 1-7 Some Familiar Red Giants
Almost every reddish star you see in the sky is a red giant. These four star maps show the names and locations of four famous red giants that are easily seen with the naked eye. Five billion years from now, the sun will become a red giant.

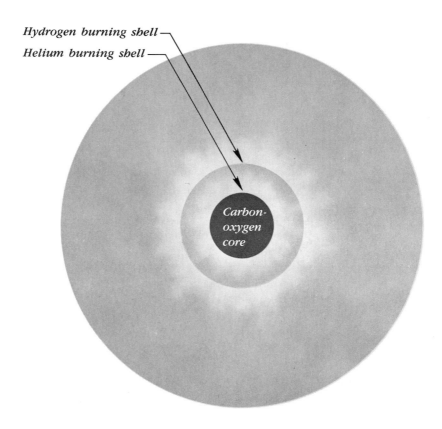

Hydrogen burning shell

Helium burning shell

Carbon-
oxygen
core

Figure 1-8   The Structure of an Old Low-Mass Star
This cross-sectional diagram shows double shell burning in a very old low-mass star. The inert core contains mostly carbon and oxygen. In low-mass stars like the sun, temperatures never get high enough to ignite any additional nuclear reactions. At this stage, the star has finished its life cycle and is just about to die. (Diagram not to scale.)

force the core temperature high enough to ignite carbon or oxygen burning. The carbon – oxygen core of the antique sun remains inert as the hydrogen and helium burning shells creep outward, greedily searching for fresh fuel.

This configuration cannot endure, and the star, becoming unstable, begins to pulsate gently. With each expansion, the star's interior cools slightly, thereby slowing the thermonuclear reactions. This decreased energy output then permits the bloated star to contract, and soon the extra compression rekindles the internal fires. This brief increase in energy output causes the star to reexpand, and the cycle starts again.

Thousands of years elapse between each of these thermal pulses. Finally the convulsions become so great that the star's outer layers separate completely from the burned-out core. These outer layers float gently into space, revealing a stellar corpse. When first exposed, the dead core may have a temperature in excess of 100,000 degrees and may shine brightly with intense ultraviolet light. This radiation causes the expanding outer layer to fluoresce and glow with unprecedented beauty. The final result is called a *planetary nebula,* surrounding the dead star like a cosmic funeral shroud.

The famous Ring Nebula, shown in Figure 1-9, is a fine example of a star that has ended its life. A dying star can eject between one-quarter and one-half of its mass in the relatively gentle process of producing a planetary nebula. Other superb examples of these funeral shrouds include the Dumbbell Nebula and the Helix Nebula, shown in Plates 5 and 6.

Planetary nebulas are short-lived phenomena in the universe. After only 50,000 years, the expanding envelope of gases has become so thinly dispersed that the nebulosity disappears from view. Meanwhile, the dead star gradually shrinks in size as it radiates its warmth into space. Unable to ignite any further thermonuclear reactions, the star contracts until it is roughly the same size as the earth. The star has become a *white dwarf.*

This shall be the final fate of our sun. Billions of years from now, long after the inner planets have been vaporized, the sun's outer layers will be expelled into space. For a brief moment, a mere 50,000 years, our solar system will be transformed into an exquisitely beauti-

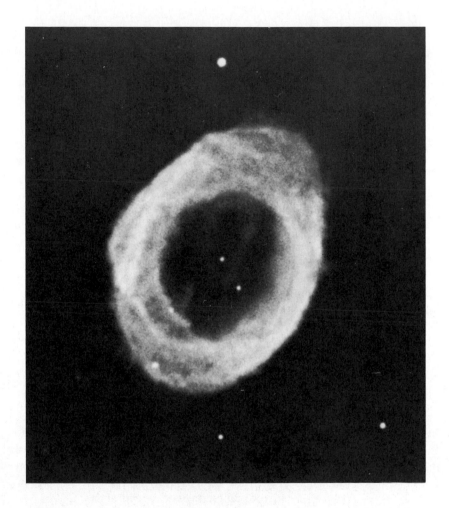

Figure 1-9   The Ring Nebula (also called M57 or NGC 6720)
This planetary nebula is one of the most beautiful sights in the sky.
About 20,000 years ago, a star was nearing the end of its life. Its
outer atmosphere was ejected, and the hot dense core was exposed.
The star exactly at the center of the ring is the corpse. (Lick
Observatory.)

19

ful planetary nebula as the shells of gas rush outward toward the darkness. The sun's burned-out core contracts and cools, begrudging release of its remaining warmth.

When the universe was created 20 billion years ago, only hydrogen and helium were formed. Only these two light elements could survive and emerge from the incredibly violent maelstrom of cosmic birth. But the world around us is composed of much heavier elements. There is oxygen and nitrogen in the air we breathe, calcium and potassium in our bones, and iron in the blood that flows in our veins.

These heavier chemicals were not part of the primordial fireball from which the universe was born. Neither were they created inside low-mass stars like the sun. Stars like the sun are incapable of creating elements heavier than carbon and oxygen. But deep inside massive stars, stars up to 50 or 60 times as massive as the sun, a host of exotic thermonuclear reactions can occur. Late in the lives of these massive stars, the tremendous weight of the star's gases pressing inward from all sides forces the central temperature to billions of degrees. In these infernos, the heavier elements are forged.

The gentle creation of a planetary nebula is far too mild for these massive stars. Instead, they end their lives in nature's most violent cataclysm: a supernova explosion. During a few brief days, these stars increase their luminosities a billionfold as they blast themselves apart. All the heavy elements manufactured deep within these stellar furnaces are then spewed across space.

Supernova explosions enrich the interstellar medium with all the heavy elements. The next generation of stars to be born from this enriched material can be accompanied by the formation of planets, moons, asteroids, and meteoroids as solid objects condense from the nebulosity surrounding contracting protostars.

This is what happened 5 billion years ago in the nebulosity that enveloped the protosun. The universe was already 15 billion years old at the time of the sun's creation. Many massive stars had already lived out their lives, so the solar nebula possessed an abundance of heavy elements. We therefore realize that every atom in our bodies, every atom we touch and breathe, was created long ago deep inside a long-forgotten star. We are literally made of the dust of stars.

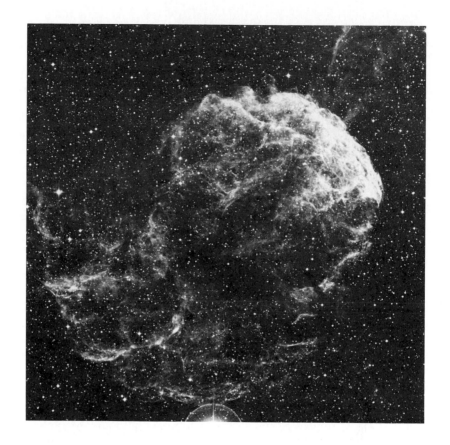

Figure 1-10   The Supernova Remnant IC 443
Tens of thousands of years ago, an ancient massive star ended its
life by becoming a supernova. The spectacular detonation blasted
enormous quantities of stellar material out into space. This
material is now joining other interstellar gases and will someday
become incorporated into new stars that are yet to be born. (Hale
Observatories.)

21

2

Stardeath and White Dwarfs

The legend of the Ephemera is most appropriate. The tale tells of an extraordinary race of insects, the Ephemera, who are blessed with great intellect but at the same time are cursed with tragically short life spans. Each of these noble insects lives for only 24 hours. During the allotted hours, each insect lives life to the fullest, partaking of the arts and sciences of the society, and making some small contribution for future generations to appreciate and enjoy.

The Ephemera live in a great forest. One of the profound mysteries facing this gifted race involved the world around them. Huge trees tower hundreds of feet into the air. Seeds and small green shoots are found on the forest floor along with numerous rotting logs.

For countless generations, the Ephemera believed that the forest is eternal and unchanging. Quite simply, none of these intelligent insects ever lived long enough to see any changes. But gradually, as aged insects told their offspring how the forest was at the time of their grandparents and great-grandparents a few days earlier, a remarkable story began to emerge. They began to realize that the objects in the forest — the trees, the sprouts, the seeds, and rotting logs — are not unrelated and unchanging. Instead the data and the reports of the elders clearly pointed to life cycles on a colossal scale. And the scientists among the Ephemera boldly theorized that the forest evolves. Seeds scattered on the forest floor germinate and turn into sprouts. Then, while thousands of generations of Ephemera come and go, the sprouts grow into gigantic trees that tower overhead. And finally, becoming old and feeble, the most ancient trees crash to the ground and by rotting enrich the soil for future generations. The forest gives the illusion of being unchanging only because the life span of each insect is so desperately short.

As you gaze out into the nighttime sky, you might first think that the heavens are eternal and unchanging. Indeed, you would be hard pressed to point out *any* changes in the stars since humanity began recording astronomical observations thousands of years ago. But during the past few decades, a remarkable scenario has been revealed. The myriad of objects and phenomena we see scattered around our Galaxy — the stars, the nebulas, the interstellar gas and dust — are not separate and unrelated. Instead they tell a remarkable story of evolution of life cycles among the stars. A star gives the illusion of being unchanging only because milestones that mark passage from infancy to adolescence or to adulthood or to old age are separated by hundreds of millions or billions of years.

As we learned in the previous chapter, starbirth can begin when a spiral arm of a galaxy moves through an interstellar cloud of gas and dust. Just as waves crashing against the seashore produce ocean spray of thousands of droplets of water, the spiral arm of a galaxy rippling across space causes enormous inter. 'ellar clouds to fragment into thousands of newborn stars. This process is easily seen in Figure 2-1, which shows details of a spiral arm of a nearby galaxy. Notice the long, arching dark lane where interstellar gases have been recently compressed and become opaque. And immediately alongside the dark lane are numerous bright nebulas where newborn stars are causing surplus interstellar gas to blaze with unprecedented beauty. As the spiral arm moves onward, the nebulosities dissipate and fade from view, leaving behind thousands upon thousands of adolescent stars.

Most young stars are very much like the sun in terms of mass, size, and chemical composition. Astronomers prefer to speak of the masses of stars in "solar masses" instead of billions upon billions of tons. By definition, the amount of matter in the sun is exactly 1 solar mass. And similarly, a star containing twice as much matter as the sun has a mass of 2 solar masses.

Scanning the skies, you occasionally find stars containing as little as one-tenth of a solar mass. These low-mass stars are quite dim. They shine with a brightness equal to only a hundredth of the sun's luminosity. With more searching, you could find high-mass stars containing 40 or 50 solar masses. These are usually the most luminous stars in the Galaxy. A young 50-solar-mass star shines with a brightness equal to a million suns.

Although it is possible to find stars that contain as little as one-tenth of a solar mass or as much as 50 solar masses, the vast majority of stars in the sky consist of roughly 1 solar mass. For example, most stars in the cluster shown in Figure 2-2 have masses between one-third and 3 solar masses.

Stellar masses are important because the mass of a star tells how fast the star evolves. Low-mass stars evolve very slowly. It takes a long time for them to build up the necessary pressures and tempera-

Figure 2-1 Details of a Spiral Arm
*The spiral arms of a galaxy are outlined by huge glowing clouds of
gas called emission nebulas. At close range, each of these nebulas
would look like the Orion Nebula, shown in Figure 1-4. Notice the
prominent, dark lanes of gas and dust that skirt the inner
(concave) edge of the spiral arm. (Hale Observatories.)*

Figure 2-2   A Cluster of Stars
*Stars can contain as little as one-tenth of a solar mass or as much
as 50 solar masses. Most stars, like those shown in this cluster, have
masses in the range of one-third to 3 solar masses. Low-mass stars
evolve very slowly, whereas high-mass stars rush through their lives
very rapidly. (Kitt Peak National Observatory.)*

tures to even begin hydrogen burning. And when thermonuclear reactions finally start, these low-mass stars consume fuel at a miserly rate. Most low-mass stars in our Galaxy are still in their childhood, still burning hydrogen bit by bit at their centers.

In sharp contrast, high-mass stars evolve very rapidly. A high-mass star easily builds up the necessary pressures and densities to ignite thermonuclear reactions. The tremendous weight of the star's outer layers pressing inward from all sides force-feeds the thermonuclear fires. Fuel is therefore consumed at a furious rate. A high-mass star can completely run through its entire life cycle even before hydrogen burning at the center of a low-mass star has been kindled.

We have seen that low-mass stars end their lives by producing planetary nebulas. When the star finds itself with an inert carbon−oxygen-rich core surrounded by shells of helium burning and hydrogen burning, instabilities develop that cause the star to pulsate. During these spasms, the dying star ejects a sizable fraction of its outer layers. For example, a 3-solar-mass star can eject nearly 2 solar masses of gas than then fluoresce and glow because of intense ultraviolet radiation from the burned-out stellar core.

But what is the nature of this burned-out stellar corpse that is left behind after the gases of the planetary nebula have dispersed and merged with the interstellar medium?

It is important to remember that thermonuclear reactions play a crucial role in the stability of a star. For example, you may recall that the ignition of hydrogen burning is responsible for stopping the contraction of a protostar. The outpouring of energy from thermonuclear reactions at a star's center sets up conditions by which the star can support the enormous weight of its outer layers.

Without any more nuclear fuel, the burned-out stellar corpse at the center of a planetary nebula simply contracts. Trillions upon trillions of tons of gas pressing inward from all sides relentlessly crush the star down to a very small size. Soon the gases are packed so tightly that atoms inside the star are completely torn apart.

Under mild conditions, such as in the everyday world around us, an atom consists of a dense nucleus orbited by electrons, as shown in Figure 2-4. Most of the atom's mass is contained in the nucleus, which is composed of particles called protons and neutrons. The protons carry positive electric charge and the neutrons are electrically neutral. The tiny, lightweight electrons that orbit the nucleus are negatively charged. Under normal conditions, the number of positively charged protons is exactly counterbalanced by an equal number of negatively charged electrons.

Deep inside a dying low-mass star, however, the atoms become packed so tightly that all the electrons are torn off their nuclei. The star's interior consists of nuclei floating in a sea of electrons. And finally, when gravity has crushed the star down to a size no bigger than the earth, the electrons are crowded together so closely that they produce a powerful pressure that vigorously resists any further squeezing. These electrons are then so closely crowded that any further contraction of the star would be like making two electrons occupy the same place (more precisely, it would be like trying to make two or more electrons occupy the same "quantum mechanical state," as physicists say). This is strictly forbidden by a law of nature called the *Pauli exclusion principle*. The resulting pressure that prevents any further contraction of the stellar corpse is called *degenerate electron pressure*.

Degenerate electron pressure can support a stellar corpse that weighs as much as 1.4 solar masses. The diameter of one of these dead stars is roughly 10,000 kilometers. The density inside the star is so great that each cubic inch of compacted stellar material weighs a thousand tons!

When the gases that form a planetary nebula lift away from a dying star, the exposed stellar corpse may have a surface temperature in excess of 100,000 degrees. The surface layers cool slightly as the star contracts. By the time the star has shrunk down to the size of the earth, the surface temperature will have dropped to about 40,000 or 50,000 degrees. The white-hot surface of one of these stars shines with a dazzling blue-white light. And because of its small size, the star is called a *white dwarf*.

White dwarfs are the most common kind of dead star in the Galaxy. All low-mass stars (including the sun) are destined to become

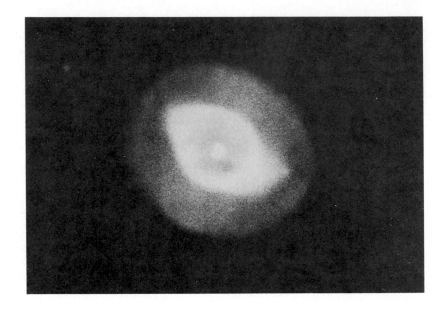

Figure 2-3   Three Planetary Nebulas
These are three typical planetary nebulas. Most planetary nebulas that we can see in the sky are roughly 20,000 years old. Their diameters can range from a few billion miles to a light year. Intense ultraviolet light from the hot central star can keep the gases glowing for about 50,000 years. (Hale Observatories.)

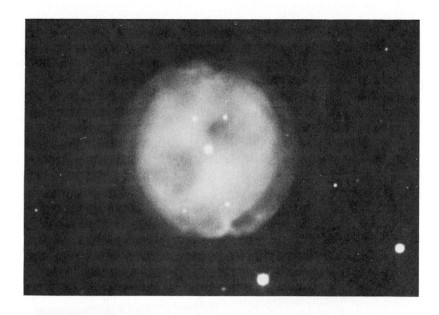

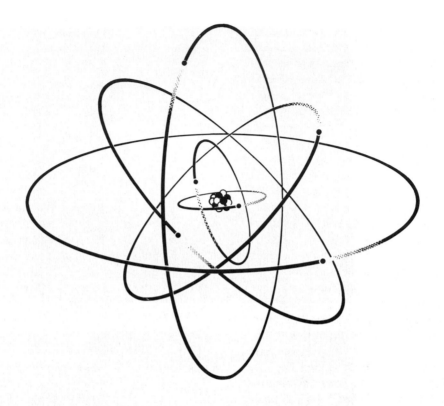

Figure 2-4   An Atom
Most of the mass of an atom is contained in the nucleus, which is composed of protons and neutrons. The nucleus is orbited by tiny, lightweight electrons. The atom is held together by the electric forces between the positively charged protons and the negatively charged electrons. Neutrons do not carry any electric charge.

white dwarfs at the end of their lives. For the rest of eternity, white dwarfs simply cool off, gradually radiating their heat into the blackness of space.

There are many white dwarfs in our Galaxy. Astronomers have discovered hundreds of them all across the sky. Sirius, for example, the brightest-appearing star in the sky, has a white dwarf companion, shown in Figure 2-5.

You may be surprised to learn that about half of the stars you see in the sky are not single stars like the sun. Instead they are double stars, like Sirius, that revolve about their common center. Some double stars have huge orbits that always keep the two stars far apart. In these cases, each star is free to live out its life relatively uninfluenced by its companion. But in many other cases, the orbits are small and the two stars are quite close together. In fact, some double stars have such small orbits about their common center that the surfaces of the two stars are actually touching! In these "close binaries," the evolution of one star can dramatically affect its companion. For example, one star may dump vast quantities of gas onto the other. Astronomers find many intriguing cases of "mass exchange" in close binary systems where there is clear evidence for streams of gas pouring off one star and crashing onto the other.

There are some fascinating double stars in the sky that consist of a white dwarf and a nearby companion that is just now becoming a red giant. Some of the surface gases of the bloated red giant fall onto the white dwarf. These gases are mostly hydrogen because they come from the red giant's outer layers, which have never been subjected to the thermonuclear fires that occur only deep within the star. Because the white dwarf is so compact (after all, typically an entire solar mass is squeezed into a volume no bigger than the earth), the surface gravity on the white dwarf is enormous. The newly arrived hydrogen-rich gases therefore weigh heavily on the white dwarf's hot surface. As more and more hydrogen accumulates on the white dwarf, pressures and temperatures in this layer of gas get higher and higher. Soon this gas is so hot that hydrogen burning explosively ignites on the white dwarf's surface! Suddenly the star increases its brightness 10,000-fold as thermonuclear reactions rage

Figure 2-5   Sirius and a White Dwarf
Sirius, the brightest-appearing star in the sky, is actually a double star. Its tiny white dwarf companion is seen in this excellent photograph. White dwarfs are very small — typically the same size as the earth — but they have surface temperatures between 10,000 and 50,000 degrees. (Courtesy of R. B. Minton.)

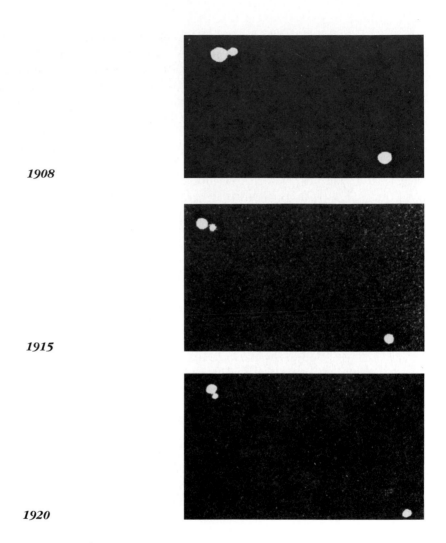

1908

1915

1920

Figure 2-6   A Double Star
About half of the stars you can see in the sky are double stars. This series of photographs spanning twelve years shows the motion of two stars about their common center. This binary star system, called Kruger 60, is in the constellation of Cepheus. (Yerkes Observatory.)

across its surface. As seen by earth-bound observers, a new star has blazed forth in the heavens. The white dwarf has become a *nova.*

Novas are a fairly common phenomenon in our Galaxy. For example, Figure 2-7 shows two photographs of a nova that erupted in the constellation of Hercules in 1935. After a couple of months, the outburst was almost finished. During an outburst, a nova can eject gases into space. The amount of material blasted away from the white dwarf is, however, quite small. Novas seldom eject more than about a ten-thousandth of a solar mass. A thin shell of ejected material about an old nova is seen in Figure 2-9.

A nova outburst is certainly a dramatic event, but it is only one of many phenomena associated with white dwarfs in close binary systems. We saw that the surface gravity of a white dwarf is very strong. This means that infalling matter from a bloated companion star crashes down on the white dwarf's surface with enormous violence. In fact, the infalling atoms are moving so fast that they emit X rays upon striking the white dwarf's surface. Astronomers have recently realized, therefore, that many newly discovered X-ray stars may actually be white dwarfs onto which gases from a nearby companion are falling.

White dwarfs were first explained in the 1930s when the Indian (now American) astrophysicist S. Chandrasekhar discovered that degenerate electron pressure could support a dead star. His calculations revealed that tightly packed electrons inside a white dwarf vigorously resist any further squeezing. This gives the star the ability to hold up the enormous weight of trillions of tons of burned-out stellar material that relentlessly press inward from all sides.

But this degenerate electron pressure is not infinitely strong. There is an upper limit to the amount of matter it can support. This important limit, called the *Chandrasekhar limit,* is 1.4 solar masses. All white dwarfs therefore must contain less than 1.4 solar masses. So what happens to a stellar corpse that contains 2 or 3 solar masses? In order to discover the fate of these stars, we must first examine the violent deaths of massive stars.

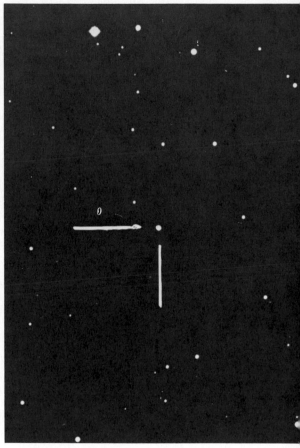

March 10, 1935

May 6, 1935

Figure 2-7   A Nova
*In 1935, a nova erupted in the constellation of Hercules. The view
on the left shows the nova at maximum brightness, just after the
outburst began. Two months later, the nova had almost completely
faded away, as shown in the view on the right. (Lick Observatory.)*

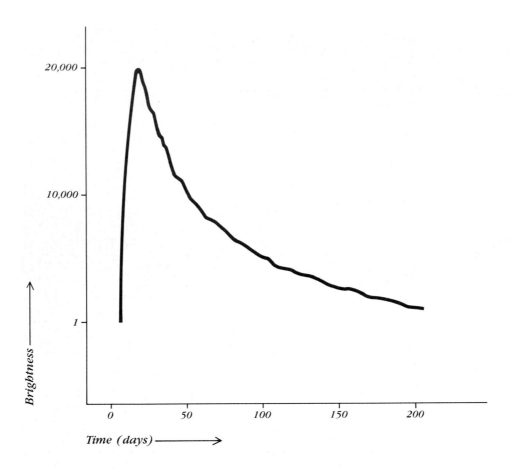

Figure 2-8   The "Light Curve" of a Typical Nova
During a nova outburst, a star flares up to thousands or even tens of thousands of times its normal luminosity. The rise to maximum brightness is very rapid, usually less than one day. The subsequent decline is much more gradual.

Figure 2-9   Expanding Nebulosity Around an Old Nova
In 1901, a nova erupted in the constellation of Perseus. This photograph was taken in 1949 and reveals an expanding shell of gas. Actually, novas eject only a small amount of matter during an outburst. (Hale Observatories.)

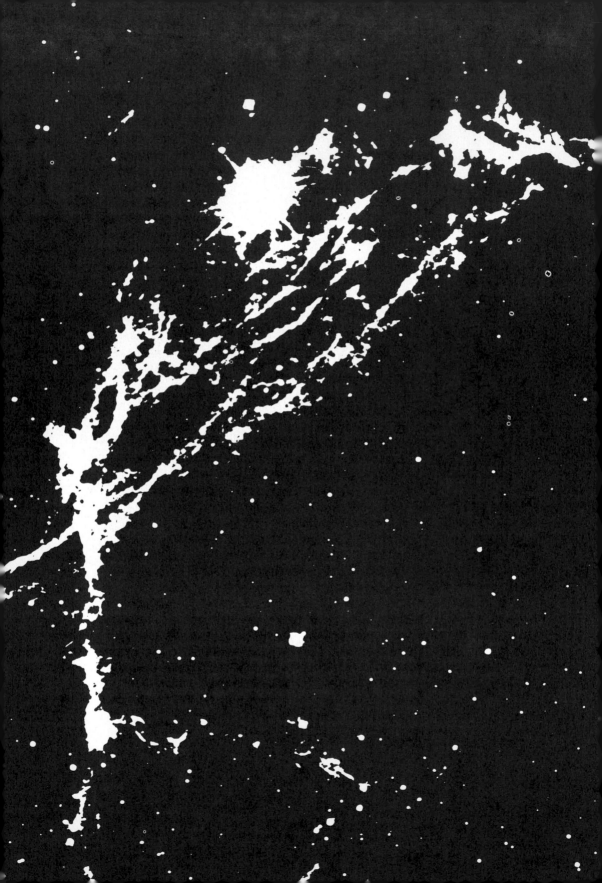

3

Supernovas
and
Neutron Stars

Most stars that you see in the nighttime sky are very much like the sun. They contain roughly the same amount of matter as the sun and are powered by hydrogen burning at their centers. Billions of years from now, after maturing into red giants, they will shed their outer layers, and their burned-out cores will contract to become white dwarfs. This is the ultimate fate of most stars, simply because most stars have fairly small masses. These stars do not have enough mass to ignite any thermonuclear reactions beyond helium burning. And with the expulsion of a planetary nebula, they usually have no difficulty ensuring that their corpses are below the Chandrasekhar limit.

Although they are in the minority, there are some dazzlingly bright stars in the sky that are considerably more massive than the sun. While still in their childhood, these massive stars are easily identified as being some of the brightest, most bluish-white stars you can find in the sky. Excellent examples include Spica (in Virgo), Achernar (in Eridanus), and almost every bright star in the constellation of Orion. Soon, however, when these stars have evolved to become red giants, it will be much harder to distinguish them from their mature low-mass cousins. All red giants have roughly the same brightness and reddish color.

Like their low-mass cousins, high-mass stars are burning both hydrogen and helium at their cores when they become red giants. But because of their enormous masses, they are entirely capable of igniting a host of additional nuclear reactions. For example, you may recall that the carbon–oxygen-rich core of a low-mass star is inert. In a high-mass star, however, the tremendous weight of overlaying stellar material forces central temperatures to 700 million degrees, and carbon burning begins. Still later, when temperatures have risen to 1 billion degrees, oxygen burning is ignited. In each case, the reaction proceeds at the star's center until all the fuel is used up. The reaction then momentarily shuts off and the star's core contracts under the influence of gravity. Soon, temperatures immediately above the contracting core are so high that the same nuclear reaction is reignited in a thin shell around the core.

The ash of oxygen burning is silicon. As a thin shell of oxygen burning moves outward, away from the star's center, an ample supply of silicon is left behind. And when further compression of the core has forced the central temperature to about 3 billion degrees, silicon burning is ignited.

Iron is the ash of silicon burning. But iron does not burn, no matter how hot the star's core becomes. Near the end of its life, therefore, a massive star has an inert, iron-rich core surrounded by several thin shells in which thermonuclear reactions are still occurring, as shown schematically in Figure 3-2. These nuclear burning shells are crowded close to the star's core. For example, in a 15-solar-mass star, all of the nuclear burning shells are within the innermost 5 solar masses. The rest of the star's gases are thinly spread out over the star's enormous, bloated volume.

The buildup of an iron-rich core signals the impending violent death of the star. Of course, the iron atoms in the burned-out core are completely torn apart. No atom is capable of surviving intact under the extreme conditions of pressure and temperature found at a star's center. The star's core therefore consists of iron nuclei floating in a sea of electrons. As the shell of silicon burning inches outward away from the star's center, more and more iron nuclei and electrons are left behind. Finally, the dead core can no longer support the crushing weight of the rest of the star. When the iron core contains about 1½ solar masses, the pressures are so enormous that the electrons are squeezed inside the iron nuclei. When this happens, the negatively charged electrons combine with the positively charged protons to form neutrons. These neutrons occupy a much smaller volume than the original iron nuclei and electrons from which they were made. Consequently, the core of the star violently implodes. This core collapse occurs suddenly and releases a vast amount of energy. In fact, the amount of energy liberated during the few brief moments of core collapse is as great as the total energy contained in all the light radiated from the star over the entire preceding course of its life! But instead of taking billions of years to gradually leak away from the star in the form of starlight, this sudden flood of energy gushes up from the star's core in only a few hours. As a shock wave from the imploding core crashes outward, the star is completely ripped apart. The star has become a *supernova.*

Ultraviolet light

Figure 3-1   Massive Young Stars in Orion
Almost every bright bluish star in the constellation of Orion is a massive young star that has recently ignited hydrogen burning at its center. These stars can be identified most easily in the ultraviolet photograph because they emit primarily ultraviolet light. (Courtesy of G. R. Carruthers, N. R. L.; Hale Observatories.)

Visible light

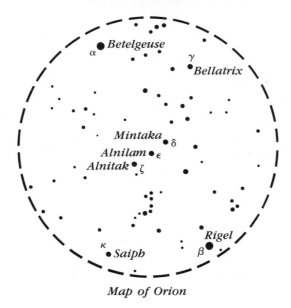

Map of Orion

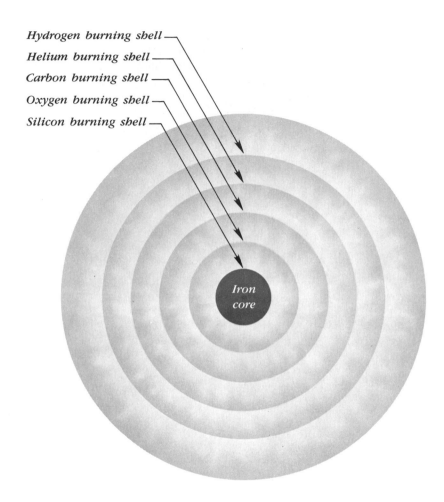

Hydrogen burning shell

Helium burning shell

Carbon burning shell

Oxygen burning shell

Silicon burning shell

Iron core

Figure 3-2   The Structure of an Old High-Mass Star
This cross-sectional diagram shows multiple shell burning inside a very old massive star. The creation of a burned-out, iron-rich core signals the impending catastrophic death of the star. A star at this final stage in its life is just about to blow itself apart and become a supernova. (Diagram not to scale.)

During a supernova explosion, a doomed star suddenly increases its brightness a hundred millionfold. For a few days, the star can actually outshine the entire galaxy in which it lived. An example is shown in Figure 3-3.

Some astronomers estimate that three or four supernovas erupt in our Galaxy every century. Unfortunately, most of our Galaxy is hidden from view by vast quantities of interstellar gas and dust. The last nearby supernova occurred in 1604 in the constellation of Serpens. The one before that exploded in 1572 in the constellation of Cassiopeia. In both cases, for a few days they outshined all the other stars in the sky. A nearby supernova has not been observed since then, and most astronomers agree that we are long overdue for a whopper.

It is important to realize that a supernova is *not* a big nova. They occur for entirely different reasons and involve entirely different kinds of stars. For example, a nova outburst can eject only a small amount of matter into space. In contrast, an enormous amount of matter can be expelled from a supernova at supersonic speeds. These ejected gases glow and fluoresce as they crash headlong into the surrounding interstellar medium. This collision can be so violent that X rays are emitted. For example, Figure 3-4 shows a beautiful supernova remnant in the constellation of Cygnus. Astronomers have recently discovered quantities of X rays coming from this expanding shell of gas as it smashes its way through the ambient interstellar gases.

We have seen how the burned-out core of a low-mass star cools and contracts to become a white dwarf as the gases of its planetary nebula dissipate and fade from view. In a similar fashion, the burned-out neutron-rich core of a high-mass star also produces a stellar corpse. But it cannot be a white dwarf. The mass is often too high, often slightly above the Chandrasekhar limit, to be supported by degenerate electron pressure.

Recall that degenerate electron pressure arose because electrons obey the Pauli exclusion principle, which explains that two electrons cannot occupy the same place at the same time. A white dwarf star is stable and does not contract anymore simply because further compression would try to force identical electrons to occupy the same place. And that is forbidden.

47

May 10, 1940   *January 2, 1941*

Figure 3-3 A Distant Supernova
*In 1940, a supernova erupted in this galaxy in the constellation
of Coma Berenices. The view on the left was taken shortly after
the outburst. Eight months later, the supernova had faded away,
as shown in the photograph on the right. At maximum brilliancy,
a supernova can be as bright as an entire galaxy. (Hale
Observatories.)*

Figure 3-4   A Supernova Remnant
This expanding shell of gas is the outer layer of a star that blasted itself apart in a supernova explosion about 50,000 years ago. A dying star can lose a large fraction of its mass in one of these cataclysms. A color photograph of a portion of this nebula is shown in Plate 7. (Hale Observatories.)

Figure 3-5   The Crab Nebula (also called M1 or NGC 1952)
*This remarkable nebula is the remnant of a supernova that
exploded in A.D. 1054. The outburst was observed and recorded by
Chinese astronomers. The nebula, located in the constellation of
Taurus, is 6,500 light years away and measures about 8 light years
in diameter. A color photograph of this nebula is shown in Plate 8.
(Lick Observatory.)*

Like electrons, neutrons also obey the Pauli exclusion principle. Two identical neutrons cannot be forced to occupy the same place. Consequently, when the neutron-rich material in the burned-out core of a massive star collapses down to a small size, *degenerate neutron pressure* arises. This powerful pressure then vigorously resists any further squeezing. The stellar corpse has become a *neutron star.*

In many respects, a neutron star is an exaggerated version of a white dwarf. A white dwarf is small and dense, but a neutron star is a lot smaller and a great deal denser. White dwarfs rotate rapidly and often have strong magnetic fields, but neutron stars rotate much more rapidly and have extraordinarily powerful magnetic fields.

We have seen that a typical white dwarf measures about 10,000 miles in diameter and is therefore about the same size as the earth. Quite simply, you must allow a solar mass of burned-out stellar matter to contract this far in order for degenerate electron pressure to arise.

A typical neutron star measures only 20 miles in diameter. Two solar masses of neutrons must collapse down to a ball 20 miles in diameter in order for degenerate neutron pressure to arise.

As you may well expect, the density inside a neutron star is absolutely enormous. So much matter is confined to such a small volume that a single tablespoon of neutron star material would weigh 40 billion tons!

Astronomers have discovered many neutron stars across the sky, although not nearly in the same abundance as the plentiful white dwarfs. These discoveries are *not* made with the usual optical telescopes found at ordinary observatories. It is virtually impossible to see an object 20 miles in diameter located among stars many light years away. Instead, astronomers identify neutron stars by searching the heavens for pulses of radio waves or bursts of X rays. These pulses of radiation are produced as a direct consequence of a neutron star's rapid rotation and powerful magnetic field.

Nearly every star you can see in the sky is rotating. In most circumstances, the rates of rotation are fairly low. For example, the sun rotates once every four weeks. But if one of these slowly rotating stars collapses down to a small size, the rate of rotation increases dramatically. The star's rotation speeds up for the same reason that an ice skater doing a pirouette speeds up when she pulls in her arms. This is why astronomers expect neutron stars to be rapidly rotating. A typical neutron star may rotate once every second or faster.

Nearly every star you can see in the sky possesses a magnetic field. In most circumstances, the strengths of these stellar magnetic fields are fairly low. For example, the sun has an overall weak magnetic field that is roughly the same intensity as the earth's natural magnetic field. But if a star possessing a weak magnetic field collapses down to a small size, the strength of the magnetic field increases dramatically. This increase occurs because the magnetic field, which originally had been spread over millions upon millions of square miles of the star's surface, becomes concentrated onto a much smaller surface area after the collapse. For this reason, we expect neutron stars to possess powerful magnetic fields. A typical neutron star may have a magnetic field a trillion times stronger than the sun's field.

In 1967, astronomers began detecting pulses of radio waves from various locations in the sky. At first these signals were very mysterious and controversial. Some astronomers even suggested that they might be signals from navigational beacons placed around the Galaxy by a race of highly advanced alien creatures! Today, however, we realize that these sources of radio waves, called *pulsars,* are actually rotating neutron stars with intense magnetic fields. When electrons on the neutron star's surface encounter the intense magnetic field at the star's north and south poles, the particles are rapidly accelerated and emit radio waves. These beams of radio waves sweep around the sky as the neutron star rotates, as diagramed in Figure 3-7. If the earth just happens to be located in the path of one of these beams, then it is possible to detect a pulse of radiation each time the neutron star rotates. The basic explanation of a pulsar is the same as of an old-fashioned lighthouse beacon.

Astronomers believe that recently created neutron stars rotate extremely rapidly and therefore give rise to the fastest pulsars. But after beaming radiation into space for thousands upon thousands of years, the neutron star gradually rotates slower and slower, and thus the pulses of radio noise become more widely spaced. Four seconds

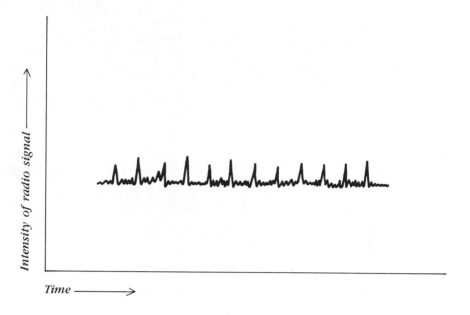

Figure 3-6   A Pulsar Recording
*In October 1967, astronomers at the Mullard Radio Astronomy
Observatory in England detected regular pulses from an object in
space. Since that time, several hundred pulsars have been found
across the sky. The fastest pulsar beeps 30 times each second. About
4 seconds elapse between pulses of the slowest pulsar. On the
average, a typical pulsar has a period around 1 second.*

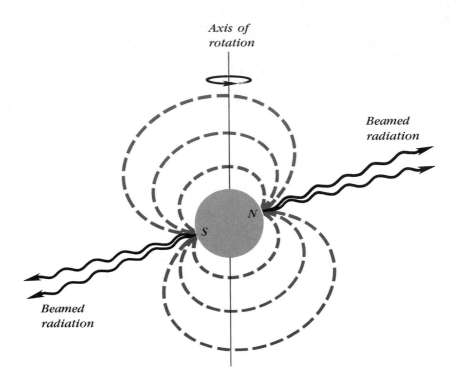

Figure 3-7   The Model of a Pulsar

Pulsars are rapidly rotating neutron stars with intense magnetic fields. Particles and radiation stream outward along narrow beams from the north and south magnetic poles. As the neutron star rotates, the beams sweep across the sky like beams from a lighthouse beacon. A person standing in the direction of the beams can see periodic pulses of radiation.

Plate 1   The Horsehead Nebula (also called NGC 2024) in Orion

Stars are born deep inside vast, cold clouds of interstellar gas and dust. These dark, frigid stellar wombs are usually very hard to find. Sometimes, however, they can be seen silhouetted against a brighter, background nebulosity or a starfield. This dark nebula, named for its characteristic shape, is roughly 1,200 light year from the earth. (Copyright by the California Institute of Technology and the Carnegie Institution of Washington. Reproduced by permission from the Hale Observatories.)

Plate 2   The Eagle Nebula (also called M16 or NGC 6611) in Serpens

Whenever a cluster of stars forms inside an interstellar cloud, the most massive stars are always first-born. These massive infant stars light up the surrounding nebulosity with their powerful ultraviolet radiation. Some dark globules, the embryos of yet-unborn stars, can also be seen silhouetted against the glowing background nebulosity, which is roughly 7,000 light years from the earth. (Copyright by the California Institute of Technology and the Carnegie Institution of Washington. Reproduced by permission from the Hale Observatories.)

**Plate 3   The Trifid Nebula (also called M20 or NGC 6514)
in Sagittarius**
Reddish nebulosity (called an emission nebula*) is caused when
hydrogen gas is forced to fluoresce by intense ultraviolet light from
newborn massive stars. Bluish nebulosity (called a* reflection
nebula*) is caused when light from cooler stars simply reflects from
surrounding interstellar gas and dust. Both types of nebulosity
shown in this photograph are located nearly 3,000 light years from
the earth. (Copyright by the California Institute of Technology and
the Carnegie Institution of Washington. Reproduced by permission
from the Hale Observatories.)*

**Plate 4   The Pleiades (also called M45 or NGC 1432)
in Taurus**
*Interstellar gas sometimes surrounds newborn stars for millions
upon millions of years. These gases will eventually be swept away
by stellar winds as the stars move from infancy to adolescence.
This young cluster can be seen with the naked eye. It is located about
400 light years from the earth. (Copyright by the California
Institute of Technology and the Carnegie Institution of Washington.
Reproduced by permission from the Hale Observatories.)*

**Plate 5   The Dumbbell Nebula (also called M27
or NGC 6853) in Vulpecula**
*The delicate funeral shroud of a planetary nebula is testimony to
the recent death of a low-mass star. The bluish stellar corpse is easily
seen at the center of the nebulosity. This dead star will soon
become a white dwarf. Meanwhile, ultraviolet light from the star's
hot surface is causing the surrounding gases to glow. This
planetary nebula is about 700 light years from the earth. (Copyright
by the California Institute of Technology and the Carnegie
Institution of Washington. Reproduced by permission from the Hale
Observatories.)*

**Plate 6   The Helix Nebula (also called NGC 7293)
in Aquarius**
*A dying low-mass star can eject half of its gases during the
relatively gentle eruption that produces a planetary nebula. As the
star's outer layers expand into space, the gases fluoresce because
of intense ultraviolet light from the exposed stellar core. After about
50,000 years, these gases are so dispersed that the nebulosity fades
from view. This planetary nebula is about 500 light years from the
earth. (Copyright by the California Institute of Technology and
the Carnegie Institution of Washington. Reproduced by permission
from the Hale Observatories.)*

Plate 7   The Veil Nebula (also called NGC 6992) in Cygnus
A massive star ends its life with nature's most violent cataclysm: a supernova explosion. The condemned star blasts itself apart, and the ejected gases glow for thousands of years as they crash headlong through the surrounding interstellar medium. Copious X rays are produced by these colliding gases. This color photograph shows only a portion of the entire supernova remnant, which can be seen in Figure 3-4. (Copyright by the California Institute of Technology and the Carnegie Institution of Washington. Reproduced by permission from the Hale Observatories.)

Plate 8   The Crab Nebula (also called M1 or NGC 1952) in Taurus

During a supernova explosion, a doomed star flares up to hundreds of millions of times its former brightness. At the peak of the outburst, the star can be as bright as an entire galaxy. All the heavy elements are manufactured in the thermonuclear inferno that tears the star apart. This supernova remnant is located 6,000 light years from the earth. The outburst that created this nebula was observed by Chinese astronomers in 1054 A.D. *(Copyright by the California Institute of Technology and the Carnegie Institution of Washington. Reproduced by permission from the Hale Observatories.)*

elapse between pulses of the slowest pulsar. This pulsar is therefore powered by one of the oldest known neutron stars in the sky.

The fastest pulsar, called the Crab Pulsar, is located at the center of the Crab Nebula in the constellation of Taurus (see Figure 3-5 and Plate 8). This dramatic nebula is the remnant of a supernova that erupted on July 4, 1054, as recorded by the ancient Chinese historian Toktaga in Part 9, Chapter 56 of *Sung Shih,* or the *History of the Sung Dynasty:* "On a chi-chhou day in the fifth month of the first year of Chih-Ho Reign-Period a 'guest star' appeared at the south-east of Thien-Kuan, measuring several inches. After more than a year, it faded away." The neutron star at the center of this nebula is therefore only 900 years old.

The Crab Pulsar is the youngest pulsar in the sky. When it was first discovered, astronomers carefully examined all the stars near the center of the Crab Nebula. It was soon noticed that one of these stars is actually flashing on and off 30 times each second. The "on" and "off" views are shown in Figure 3-8. These observations permitted the identification of a neutron star, as shown in Figure 3-9.

Inspired by the discovery of an optical pulsar in the Crab Nebula, astronomers began examining other pulsar locations for visible flashes of light. The search has proven most frustrating and very disappointing. After many years of hunting, visible flashes have been detected from only one other pulsar. This pulsar, called the Vela Pulsar because of its location in the southern constellation of Vela, is the second fastest pulsar in the sky. The Vela Pulsar beeps 11 times each second. Although its radio signals are easily picked up by radio telescopes, the corresponding visible flashes strain the limits of the finest optical telescopes in the world.

The position of the Vela Pulsar is given in Figure 3-10. The pulsar is embedded in a vast supernova remnant called the Gum Nebula. This nebulosity covers an enormous portion of the sky, an area almost 2,000 times as large as the full moon. The nebula was first examined in 1952 by the Australian astronomer Colin S. Gum, after whom it is named.

In the previous chapter, we learned that X rays can be emitted from white dwarfs in double star systems. These X rays are produced when gases from a bloated companion star crash down on the white

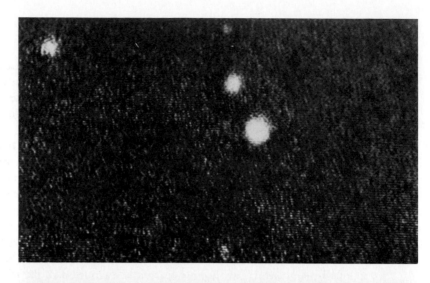

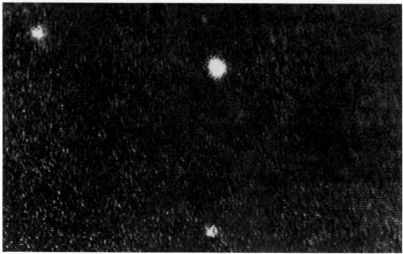

Figure 3-8 A Visible Pulsar
In 1969, astronomers succeeded in detecting visible flashes from the pulsar at the center of the Crab Nebula. The star turns on and off 30 times each second. The "on" and "off" views are shown in these two photographs. (Lick Observatory.)

dwarf's compact surface. In addition, we also saw that the accumulation of hydrogen on the surface of a white dwarf can explosively ignite, thereby producing a nova outburst. In the late 1970s, astronomers began realizing that analogous phenomena also occur with neutron stars.

In 1975, astronomers operating an X-ray-detecting satellite discovered powerful bursts of X rays coming from the central regions of a globular star cluster called NGC 6624. The bursts of X rays are roughly periodic, with the time between bursts ranging from 2 to 4 hours.

During the next few years, two dozen of these bursting X-ray sources had been discovered around the sky. They are appropriately called *bursters.* Typical X-ray data from a burster are shown graphically in Figure 3-11. For some unknown reason, a surprisingly large fraction of these bursters are located very near the centers of globular star clusters. There are many globular clusters scattered around our Galaxy. They are composed of some of the most ancient stars astronomers can find. The globular cluster shown in Figure 3-12 has a burster very near its center. Nearly one-third of all bursters are positioned at the centers of globular clusters.

Bursters will certainly be one of the exciting topics in astronomy during the 1980s. Because of their recent discovery, they are still surrounded by much controversy and confusion. Nevertheless, trends are emerging, and there is consensus on some basic issues.

Most astronomers agree that a burster is a neutron star in a double star system. Perhaps the companion star is a red giant with an enormous, bloated atmosphere. Gases from the swollen red giant are captured by the neutron star and build up on its compact surface. Soon the temperature and pressure in this material are so high that helium burning explosively ignites on the neutron star's surface. The explosion lasts for a very short time, during which a brief but powerful flood of X rays is liberated.

This explanation suggests that bursters and novas have a lot in common. Novas occur in close double star systems that contain a white dwarf. Bursters occur in double star systems that instead contain a neutron star. A nova erupts when hydrogen burning explo-

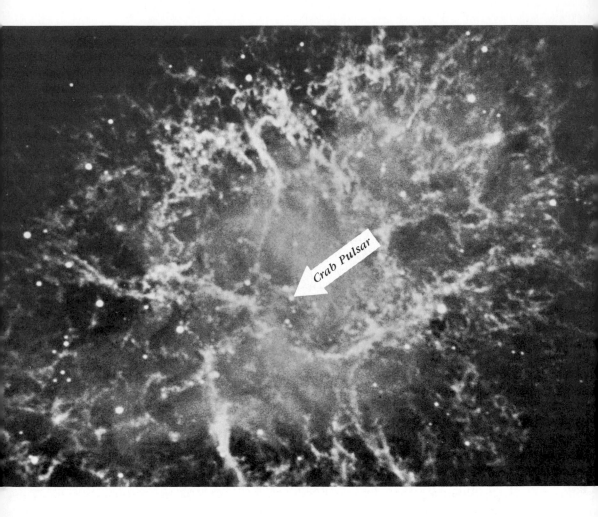

Figure 3-9   A Neutron Star
The discovery of a visible pulsar at the center of the Crab Nebula permitted the identification of a neutron star. This tiny, rapidly rotating neutron star is the powerful source of particles and radiation that have kept the nebula glowing for nine centuries. (Lick Observatory.)

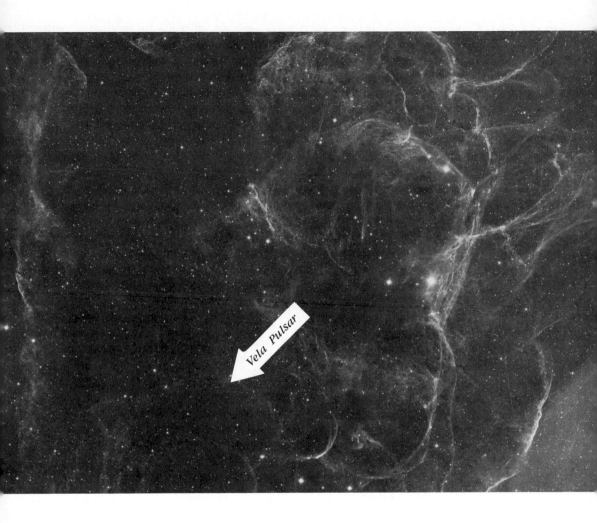

Figure 3-10   The Vela Pulsar and the Gum Nebula
*The second fastest pulsar, the Vela Pulsar, is located in the Gum
Nebula, which is a vast supernova remnant. After many years of
searching, astronomers finally detected visible pulses from this
pulsar. The pulsar is too faint, however, to be seen in this
wide-angle view. (Courtesy of B. Bok, Steward Observatory.)*

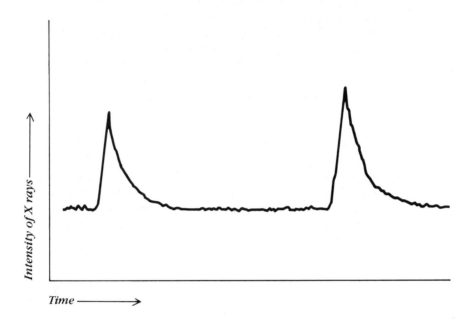

Figure 3-11   A Burster Recording
In September 1975, astronomers monitoring an X-ray-detecting satellite discovered irregular bursts of X rays from an object in space. Since that time, over two dozen bursters have been found across the sky. The time between bursts varies but typically is a few hours.

sively ignites on a white dwarf's surface. But calculations reveal that hydrogen burning on a compact star's surface is a comparatively leisurely process that continues for many days. Consequently, a nova outburst lasts for a long time. It takes months for one of these white dwarfs to return to its prenova brightness.

Hydrogen burning is far too slow to account for bursters. Astrophysicists therefore turn to helium burning for the explanation of bursters, although some scientists argue in favor of explosive carbon or oxygen burning on the neutron star's surface. Regardless of the exact fuel involved, there is a remarkable similarity between the graphs for the eruption of a burster (see Figure 3-11) and for the eruption of a nova (see Figure 2-8). And just as a burster can flare up over and over again, so can a nova. But it takes a very long time for a sufficient supply of fresh hydrogen fuel to be deposited on a white dwarf's surface. Consequently, often decades elapse between outbursts of a "recurrent nova." For example, the recurrent nova called U Scorpii flared up in 1863, in 1906, and again in 1936.

There is one final, important parallel to be drawn between white dwarfs and neutron stars. In the previous chapter we saw that there is a strict upper limit to the mass of a white dwarf. All white dwarfs must contain *less* than 1.4 solar masses. Quite simply, degenerate electron pressure is incapable of supporting more than 1.4 solar masses of burned-out stellar material.

For exactly the same reason, there is also a strict upper limit to the mass of a neutron star. Degenerate neutron pressure is incapable of supporting more than 2½ solar masses of burned-out stellar material. All neutron stars therefore must contain less than 2½ solar masses.

But there are many stars with enormous masses, stars that contain 40 or 50 times the mass of the sun. Suppose one of these massive stars fails to eject most of its gases in a supernova explosion at the end of its life. Suppose one of these stars tries to form a stellar corpse with more than 2½ solar masses of burned-out material. This is far too much matter to be supported by degenerate electron pressure, so the star certainly cannot become a white dwarf. It is also too much matter to be supported by degenerate neutron pressure, so the

Figure 3-12   A Globular Cluster
A typical globular cluster contains over 100,000 stars. This globular cluster (called M15 or NGC 7078) is about 50,000 light years away in the constellation of Pegasus. A burster is located at the center of this cluster. (Kitt Peak National Observatory.)

star will not become a pulsar or a burster. What, then, is the nature of the resulting corpse?

One of the important theoretical discoveries of the 1960s is that a massive dying star (one that contains more than 2½ solar masses) is condemned to suffer catastrophic gravitational collapse. *There are no forces in nature that can support more than 2½ solar masses of burned-out stellar material.* A massive dying star, therefore, simply shrinks under the relentless, unforgiving influence of gravity. The enormous weight of trillions upon trillions of tons of burned-out stellar material pressing inward from all sides causes the star to get smaller and smaller. Indeed, the entire star is crushed out of existence at a single point. During this implosion, gravity around the collapsing star becomes so strong that the very fabric of space and time fold over themselves, and the star literally disappears from the universe. What is left is called a *black hole.*

4

The Meaning of Warped Spacetime

For thousands of years, people have gazed up at the heavens and wondered at what they saw. Where did the stars and planets come from? Is there any meaning or purpose to the phenomena that occur across the skies?

In ancient times, people believed that the earth stood immobile at the center of the cosmos; the sun, moon, stars, and planets revolved around us, paying daily homage to our unique, central location. No wonder these people invented astrology! If we occupy a special place at the center of the universe, then it seems natural to assume that the heavens affect our lives as they revolve about us.

A fundamental lesson of the past 400 years is that these ancient ideas are totally wrong. We do not occupy a special place in the universe. We live on an ordinary planet, one of nine that orbit a typical, undistinguished star. And this star, our sun, is just one among billions scattered around our Galaxy. Even our Galaxy is commonplace. As we peer through our most powerful telescopes, we see millions of similar galaxies strewn across the depths of space.

To some people, these astronomical discoveries are depressing. To them, the lesson of modern astronomy is that humanity is a collection of insignificant microbes clinging to a small rock that orbits an ordinary star in an otherwise inconsequential galaxy — just one among billions in an inconceivably vast universe.

I prefer a different view. With each new revelation, the human mind must scale new heights and explore new dimensions. Of course, from a purely mechanical aspect, we are a race of Lilliputian life forms, poised precariously in the biosphere surrounding a very small planet. But through the human intellect, we tiny creatures have the extraordinary ability to examine and comprehend the structure of the universe. This truly distinguishes us from less-evolved animals. Not how insignificant our bodies are, but rather how potent the human mind is — this is the real lesson of modern astronomy.

A philosophical perspective based on the physical nature of reality can induce profound insights that blossom with awesome power. For example, Albert Einstein extended the non-uniqueness of our position in the universe to the very heart and soul of physics. Einstein felt that if we are learning anything of fundamental value in science, then the way in which we express our discoveries and our understanding—even the way we write down our equations—should not depend on who we happen to be, where we happen to be located, or how we happen to be moving. This was the vision of Albert Einstein. He approached physical reality with the deep personal conviction that the way in which we understand the workings of the universe should be completely independent of the position and state of motion of the observer.

This was the first time that anyone had examined physics in this philosophical light. By 1905, Albert Einstein had succeeded in reformulating everything we knew about electricity and magnetism in such a way that the equations do not depend on the exact positions or speeds of any people or observers who might do experiments and make measurements. This reformulation is called the *special theory of relativity.* But in developing this unbiased description of electricity and magnetism, Einstein found that nature had some surprises in store for us. Specifically, his theory tells us that moving clocks run slow, moving rulers shrink, and the masses of moving objects increase without bound as the speed of light is neared. The price that we must pay for a nondiscriminatory understanding of electricity and magnetism is that we must abandon many time-honored concepts and erroneous beliefs. We must cease thinking that the passage of time or the dimensions of space are permanent and inflexible. In addition, the theory teaches us to think of matter and energy as different aspects of each other, according to the famous equation $E = mc^2$. This single equation, a direct logical consequence of the equality of all observers, permitted humanity to finally understand why the sun shines. It also gave us the awesome and terrifying ability to construct thermonuclear weapons.

In the decade preceding World War I, Albert Einstein focused his attention on the phenomenon of gravity. Traditionally, everyone was taught that gravity is a *force.* According to the ideas of Sir Isaac Newton, the strength of the gravitational force between two objects depends solely on their masses and the distance between them. But Einstein had just proved that, in order for all people to have the same coherent and consistent description of reality, we must abandon our

Figure 4-1   Sir Isaac Newton (1642 – 1727)
According to Newton's theory, gravity is a force. The force of gravity holds the planets in their orbits about the sun. Using Newton's formula, scientists can calculate the orbits of planets, comets, and satellites. (Yerkes Observatory.)

Figure 4-2   Albert Einstein (1879 – 1955)
According to Einstein's theory, gravity is the curvature of spacetime. The planets follow the shortest paths in the curved spacetime about the sun. In weak gravitational fields, where spacetime is almost perfectly flat, Einstein's formulas give the same answers as Newton's. (Yerkes Observatory.)

inflexible notions about distance and mass. Consequently, we must also abandon any theory of gravity that is rooted in these erroneous notions.

It was therefore Einstein's goal to devise a description of gravity not dependent in a rigid fashion on the rulers, clocks, or scales of individual observers. He succeeded in the fall of 1915 with his formulation of the *general theory of relativity.*

The general theory of relativity is a description of gravity. It tells us how gravity works. But unlike the old-fashioned ideas of Newton, we never speak of gravity as a "force." Instead, the gravitational field around an object is manifested by curving the fabric of space and time. Far out in space, far away from any sources of gravity, space and time are perfectly flat. But as you approach a massive object, such as a star or a planet, you are moving into regions of increasingly curved spacetime. The stronger the gravitational field, the more pronounced is the curvature of spacetime.

To complete his theory, Einstein simply assumed that nature is efficient. Specifically, he argued that of all the paths that an object might follow in curved spacetime, the shortest path will always be chosen. Thus, for example, as the earth orbits the sun, it is moving along the shortest possible path through the curved spacetime that surrounds the sun. Indeed, the central idea behind general relativity is that *matter tells spacetime how to curve, and curved spacetime tells matter how to behave.*

In order to fully appreciate the theory of relativity, we must begin with a very clear idea of the meaning of *spacetime.*

Imagine looking up at the star-filled sky on a warm summer's evening. As you gaze across the heavens, your attention becomes focused on one particular star — on Vega, for example, in the constellation of Lyra, the harp. Vega is a bright bluish star that happens to be 26 light years away. This means that the light entering your eyes this particular evening has been journeying toward the earth for 26 years. You are not seeing how that star looks tonight, but rather how it appeared 26 years ago.

In exactly the same fashion, when an astronomer at a telescope photographs a galaxy that is 250 million light years away, the light that exposes the photographic emulsion has been traveling to-

Figure 4-3   The Cosmos and Spacetime
*When you look at a star that is 10 light years away, you are seeing
how that star appeared 10 years ago. When an astronomer
photographs a galaxy that is 250 million light years away, the
photograph reveals how that galaxy looked when dinosaurs
roamed the earth. Therefore, as you gaze at the heavens, you are
looking out into space and back in time. (Lick Observatory.)*

ward us for a quarter of a billion years. The photograph does not reveal how that galaxy looks now. Instead, it shows how the galaxy appeared when dinosaurs roamed the earth.

It is therefore apparent that as we gaze up at the heavens, we are looking out into space *and* back in time. By thinking about what it means to look at the stars, you are naturally led to conclude that time is a dimension to be included with the usual three dimensions of space. Indeed, if you are truly aware of what you are doing as you look up at the sky, you find that it is impossible to uniquely separate the passage of time and the dimensions of space. This four-dimensional assemblage is called *spacetime.*

We all have an intuitive understanding of the three dimensions of space. They are simply the three directions: forward – back, left – right, up – down. But, just as your ruler measures distance in the directions of space, your clock or wristwatch measures distance in the direction of time.

Scientists often find it convenient to make drawings of spacetime. These drawings are called *spacetime diagrams.* An example is shown in Figure 4-4. It is simply a graph on which time is plotted against distance.

Whenever you draw a graph, you are free to scale the axes of the graph any way you want. But scientists often find it useful to scale the axes of a spacetime diagram in a very specific fashion. If each centimeter on the vertical axis represents 1 second of time, then each centimeter on the horizontal axis shall represent 300,000 kilometers. The reason for this is that light travels at 300,000 kilometers per second (which is the same as 186,000 miles per second). The speed of light is one of the most fundamental quantities in all science. A direct consequence of the special theory of relativity is that nothing can go faster than the speed of light. Light rays on our spacetime diagram shall always travel along 45° lines because for every second that elapses, 300,000 kilometers are traversed.

Throughout this book (and in most other competent books on relativity) we are always careful to draw our spacetime diagrams so that light rays travel along lines inclined by 45° from the vertical. This has the immediate advantage of telling you where you can and cannot go in space and time. For example, examine the trip from *A*

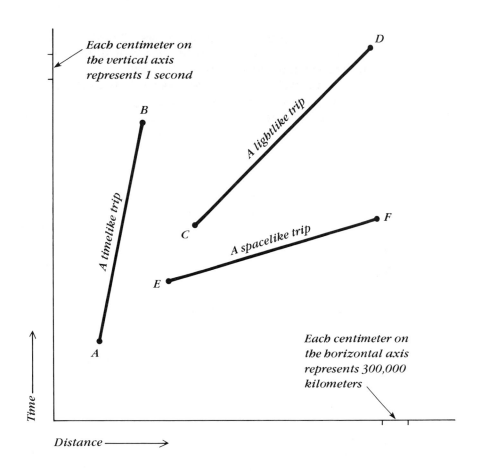

Figure 4-4   A Spacetime Diagram
A spacetime diagram is simply a graph on which time (as measured by clocks) is plotted against distance (as measured by rulers). It often proves very useful to scale the axes of the graph so that light rays travel along lines inclined at 45°.

73

to *B* in Figure 4-4. Notice that this path is inclined at an angle *less* than 45° from the vertical. To get from *A* to *B,* you leisurely move to the right as lots of time ticks by. Your speed is *less* than the speed of light. Since this journey is dominated by the passage of time, it is called a *timelike trip.*

But now look at the trip from *C* to *D.* This path is inclined by *exactly* 45° from the vertical. To travel from *C* to *D,* you would have to cover 300,000 kilometers each second. Your speed would have to be *equal* to the speed of light. This journey is therefore called a *lightlike trip.*

And finally, you could never get from *E* to *F* in Figure 4-4. The path from *E* to *F* is inclined by *more* than 45° from the vertical. You would have to cover an enormous distance in an impossibly short interval of time. Your speed would have to be *greater* than the speed of light. A journey of this type, called a *spacelike trip,* is strictly forbidden. Indeed, it can be proved that if spacelike trips were allowed, then reality would be irrational at a very fundamental level.

The basic idea behind general relativity is that gravity warps four-dimensional spacetime. Of course, visualizing curved four-dimensional spacetime would involve superhuman talents. Scientists have therefore devised some clever tricks and shortcuts to help us understand how gravity works.

Think about a star like the sun. Because of its large mass, the star is surrounded by a substantial gravitational field. For example, a person who weighs 150 pounds on earth would tip the scales at 4,200 pounds on the sun. Spacetime around the sun is more highly curved than that around the earth.

Now imagine slicing through this four-dimensional spacetime around a star and pulling out a *two-dimensional sheet of space.* Of course, you have no trouble visualizing and understanding two-dimensional surfaces. You know exactly what you mean when you say that the floor is flat, or that the surface of a football is curved. Consequently, by looking at this two-dimensional surface (technically called a *spacelike hypersurface*), you get a feeling for what gravity does to the space portion of curved four-dimensional spacetime.

This method of taking spacelike hypersurfaces can be compared to slicing through a cake to see how the layers of cake and filling are arranged. Figure 4-5 shows a two-dimensional sheet extracted from the curved four-dimensional spacetime surrounding a star like the sun. Drawings of this type are called *embedding diagrams.* Notice that far from the sun, where gravity is weak, space is flat. Strongest gravity is found immediately above the star's surface, where the curvature is most pronounced.

When Einstein first formulated his theory, he proposed an observation that might test his ideas. He reasoned that a beam of light passing near the sun's surface should be deflected from its usual straight path because the space through which it travels is itself warped. Consequently, as diagramed in Figure 4-6, the images of stars seen near the sun should be shifted slightly away from their usual locations.

Einstein's prediction was triumphantly confirmed during a total eclipse of the sun in May 1919. During the precious moments of totality, when the moon blocked out the sun's blinding disk, astronomers could photograph stars near the sun and found that they were shifted from their usual positions. Careful repeats of this and many related experiments over the next six decades leave no doubt that general relativity is by far the most accurate, complete, elegant, and precise description of gravity that humanity has ever had.

While embedding diagrams reveal the effects of gravity on the space portion of warped four-dimensional spacetime, you may be wondering what happens to the time portion. General relativity predicts that *gravity causes time to slow down.* Far out in space, far away from any sources of gravity where spacetime is flat, clocks tick at their normal rate. But as you approach a strong source of gravity, you are moving into regions of increasing gravitational curvature and your wristwatch ticks more slowly than usual. Of course, you do not notice this effect because your heartbeat, your metabolism, and even your thinking processes inside your brain have slowed down by exactly the same factor as your wristwatch. You discover this slowing of time only when you communicate with someone who remained behind in flat spacetime, where time proceeds at the normal rate.

It should be emphasized that these phenomena, the gravitational deflection of light rays by the sun and the slowing of time near an object like the earth, are exceedingly difficult to detect. You must

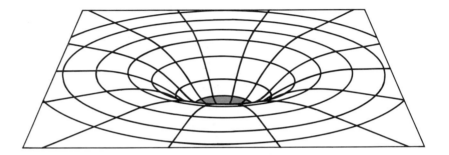

Figure 4-5   *The Gravitational Curvature of Space*
According to general relativity, gravity curves spacetime. This
drawing, called an embedding diagram, *shows how space is curved*
around a massive object such as the sun or a star. The shaded
region indicates the location of the star. The greatest curvature (and
thus the most intense part of the star's gravitational field) is found
immediately above the star's surface. Far from the star, where
gravity is weak, spacetime is almost perfectly flat.

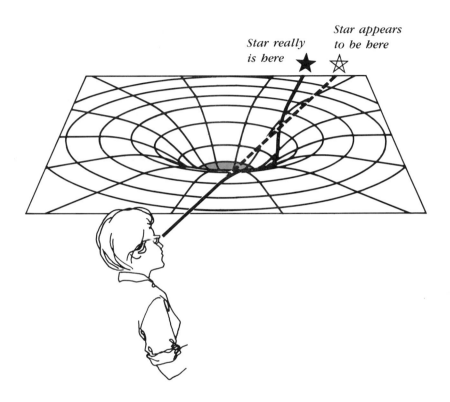

Figure 4-6   The Gravitational Deflection of Light
Light rays moving through curved spacetime must be deflected from their usual straight paths. Einstein therefore predicted that stars seen near the sun at the time of a total solar eclipse are shifted slightly from their usual positions. This deflection was first observed during an eclipse in 1919. It proved that Einstein had a better and more accurate description of gravity than Newton.

Figure 4-7   The Gravitational Curvature of Time
According to general relativity, gravity curves spacetime. While the curvature of the space portion is revealed in embedding diagrams (see Figure 4-5, for example), the curvature of the time portion is manifested in the rate at which clocks tick. Gravity slows down time. *Clocks on the ground floor of a building tick more slowly than clocks on higher floors, which are farther away from the earth's surface.*

make extremely precise measurements of star positions during a total eclipse to notice that anything unusual has happened. And it was not until 1960, five years after Einstein's death, that scientists developed clocks sensitive enough to detect the slowing of time between the ground level and upper stories of a building right here on earth.

So why bother with general relativity? Why go through the agony of complex calculations involving curved four-dimensional spacetime when the old-fashioned, seventeenth-century ideas of Isaac Newton ("gravity is a force") give excellent accuracy in almost every circumstance? And the mathematics of Newtonian gravitation are a lot simpler than Einstein's. Even when we send astronauts to the moon or launch spacecrafts toward the planets, the old Newtonian theory works superbly in calculating the orbits and trajectories.

Until recently, no one seriously believed that places might exist in the universe where spacetime is severely warped. Near the sun or a planet, around stars and galaxies, gravity is rather weak and spacetime is only slightly curved. That is why the old-fashioned Newtonian ideas work so well in so many circumstances. In weak gravitational fields, it is reasonable to replace the effects of curved spacetime with the effects of a force.

In the 1960s, astronomers finally began making major advances in understanding the life cycles of stars. This understanding clearly demonstrated that massive dying stars catastrophically implode under the overwhelming influence of gravity. Gravity around one of these massive dying stars is no longer weak. Indeed, the curvature of spacetime becomes so severe that the doomed star separates from our universe, leaving behind a hole in the cosmos.

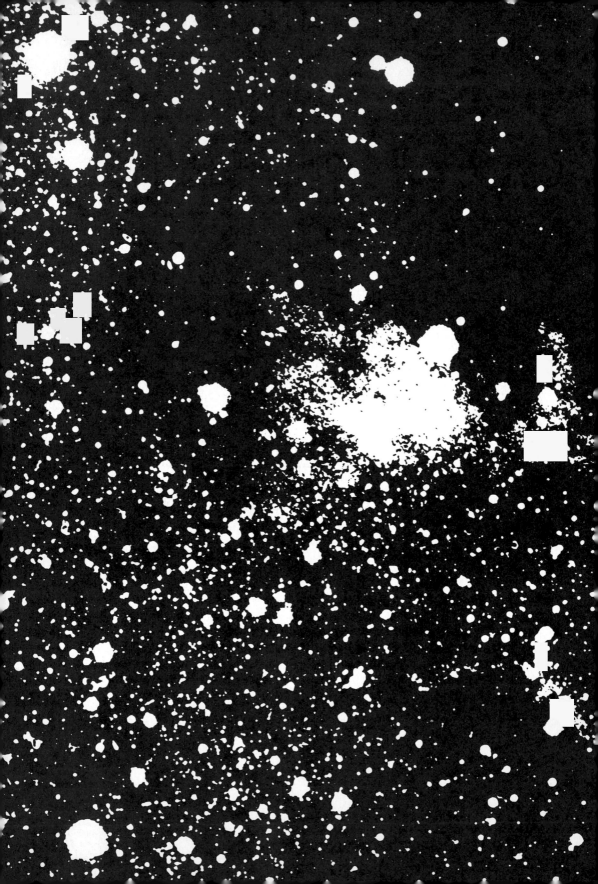

5

The Structure
of a Black Hole

Imagine a massive star at the end of its life. All internal sources of thermonuclear fuel have been depleted. A supernova explosion has just ripped the star apart. Although the star has violently shed most of its gases, the remaining burned-out core contains more than 2½ solar masses. No forces in nature can support this stellar corpse; it is doomed to become a black hole.

Before the gravitational collapse begins, gravity around the burned-out star is still comparatively weak. Of course, a person standing on this star would weigh several tons. But this condition is negligible in the face of what soon occurs. In terms of general relativity, spacetime around the condemned star is still only slightly curved. Light rays leaving the star are only slightly deflected from their usual straight-line paths.

The collapse proceeds rapidly as soon as gravity starts to overwhelm the forces between the particles inside the burned-out star. In a matter of seconds, the star dramatically shrinks in size as these particles — the electrons, protons, and neutrons — are literally squeezed inside each other. As gravity compresses the star's material into a smaller and smaller volume, the curvature of spacetime around the star becomes increasingly pronounced. Because of this increasing curvature, light rays leaving the star's surface are deflected through larger and larger angles. In fact, as diagrammed in Figure 5-1, some of the light rays that normally would have escaped into space are now pulled back to the star's surface.

As the gravitational collapse races toward its inevitable fate, more and more light rays are bent back down toward the shrinking star's surface. As the curvature of spacetime grows, only those light rays leaving the star near the vertical still manage to escape. Because more and more light is pulled back to the star, distant astronomers would see the star rapidly growing darker and darker.

Finally, at a critical stage in the collapse, the curvature of spacetime becomes so great that *all* light rays get bent back to the shrinking star's surface. Even light rays that leave the star in the vertical direction are pulled down. Since all light leaving the star — regardless of direction — is deflected back by severely curved spacetime, the

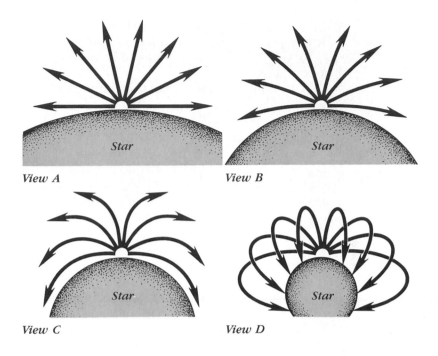

View A

View B

View C

View D

Figure 5-1   Light Rays from a Collapsing Star
As a massive dead star collapses, the intensity of gravity above the star gets stronger and stronger. This means that, as the collapse proceeds, light rays leaving the star's surface are bent through increasingly larger angles. View A shows the star just before the onset of collapse; spacetime around the star is only slightly curved, and light rays leave the star along nearly straight lines. Views B and C show the star during collapse; as the curvature of spacetime around the star increases, light rays become increasingly deflected from their usual straight paths. This deflection eventually becomes so severe that, as shown in view D, all light rays are bent back down to the collapsing star's surface. At this stage the star has fallen inside its own event horizon, and a black hole has formed.

star has become black. And since nothing can travel faster than light, nothing else can manage to escape from the star to the outside universe. Gravity has become so strong that, quite literally, the star has disappeared from the universe.

When the collapsing star has contracted to the stage that nothing — not even light — can escape, we say that the star has fallen inside its *event horizon*. For example, when a 10-solar-mass dead star has shrunk down to a sphere 60 kilometers (37 miles) in diameter, the star disappears inside its event horizon. It disappears because gravity has become so strong that light rays from the star can no longer reach the eyes of anyone watching the implosion from a safe distance. We therefore say that the diameter of a 10-solar-mass black hole is 60 kilometers (37 miles).

The term "event horizon" is very appropriate. It is literally a *horizon* in the geometry of space and time beyond which you cannot see any *events*. You have no way of knowing what is happening inside an event horizon. You cannot communicate with anyone on the other side of an event horizon. It is a place disconnected from our space and time. It is not part of our universe.

The diameter of the event horizon is simply proportional to the mass of the dead star. For example, a 5-solar-mass stellar corpse has an event horizon whose diameter is 30 kilometers (18½ miles). And the diameter of the event horizon of a 20-solar-mass black hole is 120 kilometers (74 miles). The graph in Figure 5-2 relates the masses of black holes and the diameters of their event horizons.

As seen by a distant astronomer, a black hole has formed once a dying star has shrunk inside its event horizon. But there are still no forces in nature that can support the star. So it continues to contract under the relentless influence of ever-increasing gravity. The strength of gravity and the curvature of spacetime around the imploding star continue to grow until the entire star is crushed down to a *single point!* At that point there is infinite pressure, infinite density, and, most importantly, infinite curvature of spacetime. This is where the star goes. Every atom and every particle in the star is completely crushed out of existence at this place of infinite spacetime curvature. This is the heart of a black hole. It is called the *singularity.*

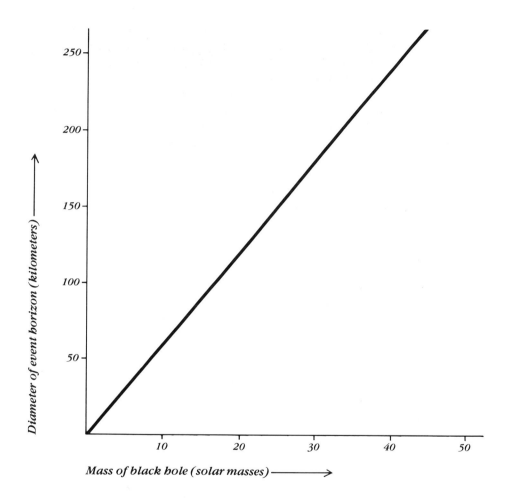

Figure 5-2 The Diameter of the Event Horizon
The diameter of the event horizon around a static black hole is directly related to the mass of the hole. The bigger the mass, the bigger the event horizon. (This graph is for black holes that are uncharged and not rotating.)

It is interesting to note that black holes are very simple. As shown in Figure 5-3, our black hole has only two parts: a singularity surrounded by an event horizon. And that's all!

It is also important to realize that the black hole is *empty.* *There is absolutely nothing there!* No atoms, no rocks, no gas, no dust. *Nothing!* Although scientists often refer to the event horizon as the "surface of a black hole," there is nothing physical or tangible at the event horizon. All the star's matter has been completely crushed into the singularity at the hole's center. The black hole is completely empty. All that exists is a region of highly warped space and time.

Many of the strange effects of general relativity — the same effects that are so infinitesimal here on earth or near the sun — are exaggerated beyond all believable bounds near a black hole. For example, we saw that gravity slows down time (recall Figure 4-7). While this phenomenon is entirely negligible here on earth, *time stops completely at the event horizon* surrounding a black hole. If you watch as your friend plunges toward a black hole, you see his clocks running slower and slower. And at the moment of piercing the event horizon, the infalling clocks appear to stop completely as time forever stands still. Of course, your ill-fated friend notices none of this. All of his bodily functions — his heartbeat, his metabolism — and any other time-keeping processes are all affected to exactly the same extent as his wristwatch is. Your friend feels himself plunge through the event horizon at nearly the speed of light on a fatal journey to infinitely curved spacetime.

Inside the event horizon, we find that the directions of space and time become interchanged! To see what this means, think about your own life on earth. You have complete freedom to move in any of the three space directions: up – down, left – right, and forward – back. But whether you like it or not, you are dragged along through the time direction from the cradle to the grave. Inside a black hole, you do indeed have freedom to move through time, but it does you no good. Whatever you gain in freedom with time, you lose in freedom with one of the directions of space. You find that you are hopelessly dragged along the space direction straight into the singularity.

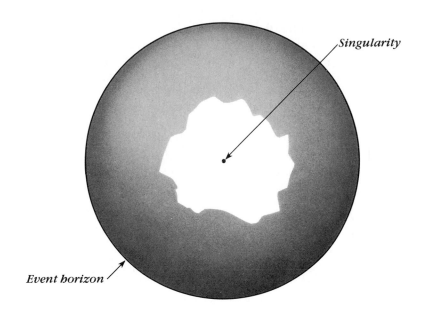

Figure 5-3   The Structure of a Static Black Hole
*A black hole consists of an event horizon surrounding a singu-
larity. At the event horizon, gravity is so strong that nothing — not
even light — can escape from the hole. Inside the event horizon, the
intensity of gravity and curvature of spacetime continue to increase
right up to the singularity. At the singularity, there is infinite
curvature of spacetime.*

Until now we have focused our attention on the simplest possible type of black hole: a black hole that contains only matter. Specifically, the hole does not carry any electric charge. And it is not rotating. This uncharged, static black hole consists of a solitary event horizon surrounding the singularity. It is sometimes called a *Schwarzschild black hole* because the equations describing the hole were first written down by the German astronomer Karl Schwarzschild in 1916.

Black holes obviously must contain matter. Without the gravity that accompanies matter a black hole could never form in the first place. Therefore the *mass* of a black hole (that is, the total amount of matter that went to the singularity) is an important characteristic of the hole. Indeed, we learned from Figure 5-2 that the mass of the hole determines the diameter of its event horizon.

Are there any other properties or characteristics that a black hole can have?

It is important to realize that black holes eat stuff in an unforgiving, irreversible fashion. An object that falls into a black hole is forever removed from our universe. Since the object is no longer part of our universe, many of its properties are no longer detectable. Therefore these properties cannot influence the structure of the hole.

This important point is perhaps best illustrated by an example. Suppose you make two black holes. One black hole is made from 10 solar masses of peanut butter while the other black hole is made from 10 solar masses of bricks. Before gravitational collapse, you have no trouble telling which is which. Peanut butter looks, smells, tastes, and feels very different from bricks. But after gravitational collapse, it is impossible to tell the two holes apart. Because the infalling matter is separated from us by an event horizon, we have no way of discovering which hole ate the bricks and which ate the peanut butter. Since many of the properties of infalling matter are swallowed by the hole, these properties (for example, color, texture, and chemical composition) cannot influence the structure of the hole.

Although individual properties of infalling matter cannot affect the structure of a black hole, we *can* know the total amount of matter that has been swallowed by the hole. We can measure the mass of the hole by placing a satellite in orbit around the hole. By carefully observing the satellite's orbit we can calculate the total amount of matter inside the hole. The answer, a number that gives the hole's mass, is one of only three numbers on which the hole's structure can depend. We can measure the mass of a black hole because gravity is a long-range force. The gravity carried by the hole's mass can be felt over a vast distance.

Like gravity, electricity is also a long-range force. Therefore, if a black hole carries an electric charge, the resulting electric field can be detected and measured over a vast distance. With appropriate equipment, an orbiting satellite can be used to measure the total amount of electric charge possessed by the hole. Since this property of a black hole can be measured by people far from the hole, the hole's structure must be influenced by its total electric charge. The total electric charge possessed by a black hole is the second of the three numbers on which the hole's structure can depend.

Imagine an ordinary, uncharged, static black hole such as the one in Figure 5-3. Now suppose that you begin giving the hole an electric charge, perhaps by dropping in some electrons. As the hole begins to develop an electric charge, a *second event horizon* forms around the singularity. Since the hole has two properties (mass and charge), it now has two event horizons. There are now two places around the singularity where time appears to stop. An electrically charged black hole would therefore have the structure shown in Figure 5-4.

As more and more electric charge is added to the hole, the inner event horizon gets bigger and bigger, while the outer event horizon shrinks. The maximum possible charge on a black hole occurs when the two event horizons finally merge. If you tried to force the black hole to accept more charge, both event horizons would disappear, leaving a *naked singularity*. These properties were first

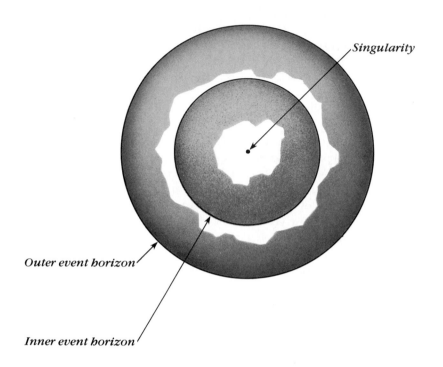

Figure 5-4   The Structure of an Electrically Charged Black Hole
An electrically charged black hole has a second (inner) event hori-
zon. With the addition of more and more charge, the inner event
horizon grows while the outer event horizon shrinks. The maximum
possible charge occurs when the inner and outer event horizons
merge.

formulated between 1916 and 1918 by H. Reissner in Germany and G. Nordstrøm in Denmark. An electrically charged black hole is therefore sometimes called a *Reissner-Nordstrøm black hole.*

Although it is theoretically possible to have electrically charged black holes, we do not realistically expect to find any in space. The reason is that, ounce for ounce, electric fields are enormously more powerful than gravitational fields. For example, imagine two electrons separated by a few centimeters. The electric force between these two particles is a trillion trillion trillion times as strong as the gravitational force that they exert on each other. Consequently, if a black hole possessed an electric charge, it would be surrounded by an enormously powerful electric field. This electric field stretching outward for many light years in all directions would easily tear apart atoms of the surrounding interstellar gas and dust. If the hole originally possessed a positive charge, vast numbers of negatively charged electrons from surrounding regions would be vigorously attracted to the hole by its electric field. In this way, the hole would soon become electrically neutral.

While we do not expect to find any electrically charged black holes around our Galaxy, we would indeed expect real black holes to be rotating. As we saw in Chapter 3, virtually all stars in the sky are rotating. These rotation rates are often quite low. But if one of these stars collapses to form a black hole, its speed of rotation increases dramatically. We therefore realistically expect rotation to be an important property of black holes formed from the corpses of massive dead stars. Although they probably would not have any electric charge, they most certainly would possess spin.

Rotation is the third and final property that a black hole can possess. Like mass and charge, the spin of a hole can be measured by scientists at safe distances from the hole. This is possible because a rotating hole literally drags space and time around itself. If you shine beams of light past a rotating black hole, the beams are deflected by different amounts, depending on whether the beams are going upstream or downstream in the spacetime rotating with the hole. In fact, by measuring the deflection of light on either side of a rotating

Figure 5-5   Interstellar Gas and Dust
*Space between the stars is not an absolutely perfect vacuum. If an
electrically charged black hole existed somewhere in our Galaxy,
its powerful electric field would tear apart atoms of interstellar gas
and dust for many light years in all directions. The hole's electric
field would then vigorously attract many particles whose electric
charge is opposite to the charge on the hole. A negatively charged
hole, for example, would attract vast numbers of positively charged
protons. Therefore, even if an electrically charged black hole
existed, it would soon neutralize itself. (Lick Observatory.)*

black hole, you can deduce the total spin of the hole. Spin is the third of only three numbers on which the structure of a black hole can depend.

Imagine a simple, uncharged, static black hole such as the one in Figure 5-3. If you start the hole spinning, a *second event horizon* forms around the singularity. Since the hole has two properties (mass and spin), it now has two event horizons. Just as in the case of electrically charged black holes, there are again two places around the singularity where time appears to stop.

If you increase the hole's rotation rate, the inner event horizon grows bigger and bigger while the outer event horizon shrinks. The maximum possible spin of a black hole occurs when the two event horizons merge. If you tried to make the hole rotate faster, both event horizons would disappear, leaving a naked singularity.

Although this description of a rotating black hole parallels our discussion of an electrically charged black hole, there are some very significant differences. In both static and charged black holes, the singularity is a *point*. As shown in Figure 5-6, the singularity of a rotating black hole is a *ring*. This ring singularity lies in the plane of the hole's equator, perpendicular to the hole's axis of rotation.

In a static or charged black hole, anyone who plunges toward the center of the hole gets torn apart by infinitely curved spacetime. Regardless of the direction you approach a point singularity, you are doomed.

In a rotating black hole, however, you are doomed to strike the singularity only if you approach the ring singularity edge-on. You get torn apart by infinitely curved spacetime only if you plunge toward the hole's center in the plane of the hole's equator. If you approach the hole's center at any other angle, you go *through* the ring without encountering infinitely curved spacetime! Going through the ring singularity is not like going through a hole in a doughnut; you do not simply come out the other side. Instead, as we shall see in the next chapter, you enter *negative space,* where gravity is repulsive. In moving through the ring singularity, you enter an antigravity universe, a place where gravity pushes things *up* rather than pulls things down!

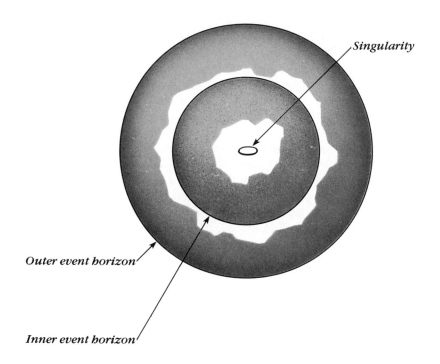

Figure 5-6 The Structure of a Rotating Black Hole
*A rotating black hole has a second (inner) event horizon. If the
hole's rotation rate speeds up, the inner event horizon grows while
the outer event horizon shrinks. The maximum amount of spin is
achieved when the two event horizons merge. Unlike the two other
types of black holes, a rotating black hole has a ring-shaped
singularity. The ring singularity lies in the plane of the hole's
equator, perpendicular to the hole's axis of rotation.*

The Structure of a Black Hole

The dragging of spacetime around a rotating hole and the surprising nature of the ring singularity were two items clarified by the work of the New Zealand mathematician Roy P. Kerr. Working at the University of Texas in 1963, Dr. Kerr was the first person to succeed in writing down the equations that completely describe a black hole with both mass and spin. Rotating black holes are therefore sometimes called *Kerr black holes*. They are the kinds of objects that we realistically expect to find scattered around our Galaxy as the collapsed corpses of massive stars.

6

Exotic Properties of Black Holes

Black holes are among the simplest objects in the universe. At most, the structure of a black hole is completely determined by only three numbers: the hole's mass, charge, and spin. Other objects in the universe — such as people, planets, and stars — are considerably more complicated. For example, to completely describe the structure of the sun, you would have to supply vast quantities of data about the pressure, density, temperature, and chemical composition of the solar gases ranging from the sun's center to its edge. None of these details enters into the complete description of a black hole.

As we saw in Chapter 4, embedding diagrams are one of the powerful tools used in understanding general relativity (recall, for example, Figures 4-5 and 4-6). By slicing through warped four-dimensional spacetime and extracting a two-dimensional sheet of space, we can actually see what gravity does to space. Perhaps, therefore, we would gain important insight into the nature of a black hole by examining embedding diagrams of a collapsing massive star.

Imagine a massive star at the end of its life. For simplicity, suppose that the star is uncharged and not rotating. It is doomed to become a static (Schwarzschild) black hole.

Prior to the onset of collapse, an embedding diagram of the space around the massive star simply looks like an exaggerated version of an embedding diagram of the space around our sun. As shown in Figure 6-1, the embedding diagram of the massive star has a deeper depression than the embedding diagram of the sun (compare with Figure 4-5). After all, the condemned massive star contains considerably more matter than the sun.

As the collapse proceeds, gravity around the imploding star becomes stronger and stronger. The curvature of spacetime becomes increasingly pronounced, and therefore the depression in the embedding diagram gets deeper and deeper. The final form of the embedding diagram following the creation of the black hole was first investigated by Einstein and Rosen in the 1930s. Much to their surprise, they found that the embedding diagram opens up and connects to another universe! As shown in Figure 6-2, the embedding diagram of a static black hole connects our universe (the upper sheet) to a second, parallel universe (the lower sheet).

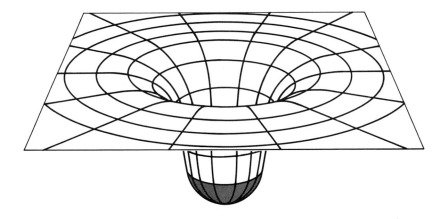

Figure 6-1   The Embedding Diagram of a Massive Star
*The curvature of space around a massive star is clearly displayed
in an embedding diagram. Far from the star, space is flat because
gravity is weak. The strongest gravity and the most pronounced
curvature of space are found immediately above the star's surface.
The shaded region indicates the location of the star's matter.*

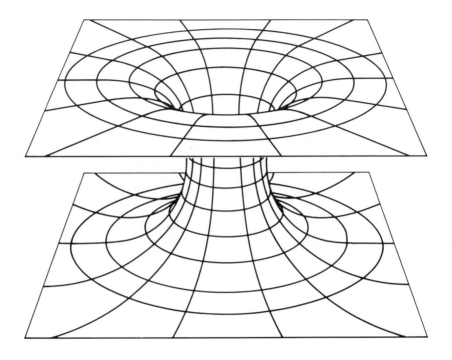

Figure 6-2   The Embedding Diagram of a Static Black Hole
*The embedding diagram of a static black hole connects our
universe (the upper sheet) with a second, parallel universe (the
lower sheet).*

This extraordinary geometrical property of a static black hole is called an *Einstein-Rosen bridge.* Quite simply, if you go up to a static black hole and slice through it in order to pull out a sheet of space, you find that a second sheet is attached to your original sheet.

The concept of a second, parallel universe is only one possible interpretation of the lower sheet of the Einstein-Rosen bridge. An equally acceptable explanation is that this lower sheet could be a remote portion of our own universe. This is possible because the upper sheet might actually be connected to the lower sheet far from the hole, as shown in Figure 6-3.

To complete this alternative interpretation of the Einstein-Rosen bridge, we can simply unfold the geometrical shape in Figure 6-3. In this way, we ensure that our universe far from any black holes looks as flat as it really is. The final result, shown in Figure 6-4, is called a *wormhole.*

The Einstein-Rosen bridge actually turned out to be only the tip of the iceberg. During the 1950s, mathematicians began discovering that a simple-minded approach to the structure of black holes has many severe problems. Specifically, if you just think of black holes as spherical event horizons surrounding singularities (like Figures 5-3, 5-4, and 5-6), you can run into some big problems with your computations. For example, a complete calculation describing someone falling into a static black hole predicts that at the event horizon the person can meet himself or herself traveling backward in time, rising up out of the singularity.

These bizarre results tell us that we still do not have a complete understanding of black holes. Something is missing. There must be more to the geometry of black holes.

A major breakthrough occurred in 1960 with the work of M. D. Kruskal and G. Szekeres. They showed that problems arose with calculations about static black holes because simple-minded diagrams like Figure 5-3 have two complete sets of spacetime piled on top of each other. If you instead unfold the overlapping regions of space and time and lay them out in a spacetime diagram, you eliminate all the bizarre problems. Infalling people are never in danger of meeting themselves going backward in time.

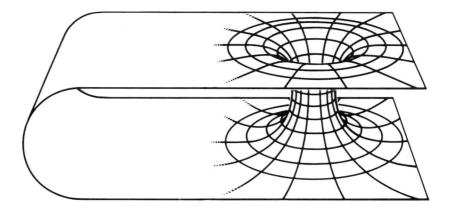

***Figure 6-3   The Einstein-Rosen Bridge (An Alternative
Interpretation)***
*The two sheets of an Einstein-Rosen bridge can be interpreted as
different portions of our own universe. This is possible if the upper
sheet actually happens to be connected to the lower sheet far from
the black hole.*

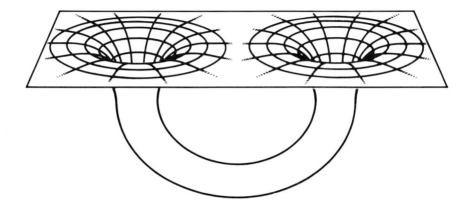

Figure 6-4   A Wormhole in Spacetime
*The spacetime of our unvierse is almost perfectly flat. This is
because there is very little gravity in the space between stars and
galaxies. Consequently, any curvature of space and time between
widely separated stars or galaxies must be very slight. It is therefore
proper to redraw Figure 6-3 so that our universe does not have any
folds or bends.*

Perhaps this is not really as strange and surprising as it first sounds. After all, general relativity speaks of *spacetime.* In relativity theory, we learn to treat time on an equal footing with space. Therefore, to really understand black holes, we should not rely exclusively on drawings in space alone (such as Figures 5-3, 5-4, and 5-6). Neither should we rely on the spacelike hypersurfaces of embedding diagrams. Embedding diagrams like Figures 6-2, 6-3, and 6-4 show only those portions of space that are attached to the flat space far from the hole. None of these diagrams shows any space inside the event horizon. They do not even show the singularity. We really need a complete spacetime diagram. We want spacetime maps that properly display all the regions of space and time associated with black holes.

In the late 1960s, the British mathematician Roger Penrose developed a very powerful method of drawing maps that properly display *all* regions of spacetime associated with black holes. In order to understand these maps, which are often called *Penrose diagrams,* we must first remember some important characteristics of spacetime diagrams in general.

In spacetime diagrams (recall, for example, Figure 4-4), time is measured vertically while distances through space are plotted horizontally. Also remember that we are careful to scale our diagrams so that light rays travel along 45° lines, as demonstrated in Figure 4-4. This has the advantage of immediately telling us where we can or cannot go. Specifically, we can always travel along any path that is inclined by less than 45° from the vertical. Along such paths (called timelike trips), our speed is always less than the speed of light. On the other hand, paths inclined by more than 45° from the vertical are always forbidden. Such paths (called spacelike trips) would require faster-than-light travel.

The Penrose diagram of a simple, static black hole is displayed in Figure 6-5. Every square inch of space and time associated with the hole is laid out before you. There are no overlapping regions; all the various portions of space and time are properly sorted out, as required by the discoveries of Kruskal and Szekeres.

The first thing we notice about the spacetime map in Figure 6-5 is that our universe is lined up along the left side of the diagram.

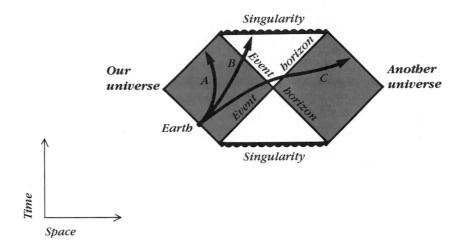

Figure 6-5   *The Spacetime Map of a Static Black Hole*
This map shows all of the regions of space and time associated with a static (Schwarzschild) black hole. Our universe is located on the left side of the map. The other universe (the second sheet of the Einstein-Rosen bridge) is on the right side. Trips A and B are allowed. Trip C is forbidden because it requires traveling faster than the speed of light.

105

The left edge of the diagram is the same as the upper sheet of the Einstein-Rosen bridge. Even the position of the earth is identified.

We also notice that the "other universe" associated with a static black hole is lined up along the right side of the diagram. The right edge of the diagram is the same as the lower sheet of the Einstein-Rosen bridge.

Remember that simple-minded drawings of black holes (such as Figure 5-3) present problems because several distinct regions of space and time are actually piled on top of each other. Just as Kruskal and Szekeres found that there are really two outside universes associated with a static black hole, they also found that there are really two singularities. One singularity lies in the past (at the bottom of the Penrose diagram) and one is located in the future (at the top of the Penrose diagram). Of course, the "past singularity" does not bother us too much. Since no one knows how to travel backward in time, you could never get there. But when you fall into a static black hole, you are condemned to strike the "future singularity" at the top of the diagram.

Finally, event horizons crisscross the Penrose diagram, thereby separating the two outside universes from the two singularities. The event horizons are inclined at exactly 45° because (in the best traditions of Alice in Wonderland!) you would have to be traveling outward at the speed of light in order to stand still on the event horizon. To understand why this is so, imagine trying to hover somewhere above the black hole. Obviously you need a set of retrorockets. Far from the hole, small rockets will suffice. But if you want to hover close to the hole, your rockets would have to be extremely powerful to prevent you from falling in. In fact, for you to hover at the event horizon, your rockets would have to be blasting so furiously that, if you were in free space, you would essentially be moving at the speed of light.

By drawing Penrose diagrams such as Figure 6-5, we immediately see where we can or cannot go. For example, consider trip *A* in Figure 6-5. An astronaut following this path leaves the earth on a journey toward the hole. But before getting near the event horizon, this astronaut gets cold feet and steers away from the hole. She ends up somewhere else in our universe. In spite of the astronaut's faintheartedness, trip *A* is a perfectly allowable journey. Nowhere is the path inclined by more than 45° from the vertical.

Trip *B* is also an allowable journey that is everywhere inclined by less than 45° from the vertical. But someone who follows such a path is surely a suicidal maniac! Once you pass through the event horizon, you are doomed to strike the singularity. Since the event horizons in Penrose diagrams are always inclined at exactly 45°, once you pass through an event horizon, there is no way of getting back out. Since you are destined to move along paths inclined by less than 45°, it is impossible for you to return to your home universe.

Of course, trip *C* is forbidden. The path is inclined by more than 45° from the vertical and therefore would require faster-than-light travel. Indeed, a moment's thought about the Penrose diagram in Figure 6-5 reveals that any path from "our universe" to the "other universe" must have some spacelike portions. It is therefore impossible to cross from one side of the diagram to the other. It is impossible to get from one sheet of the Einstein-Rosen bridge to the other. You cannot get through the wormhole.

Incidentally, we can now understand why the Einstein-Rosen bridge looks the way it does. Imagine slicing horizontally straight across the middle of the Penrose diagram in Figure 6-5. You then examine this spacelike slice that extends from our universe directly across to the other universe. Distant regions of these two universes lie far from the black hole. Far from the hole, gravity is weak and space is flat. Consequently, your embedding diagram must have two sheets that become quite flat far from the hole. As you approach the hole from either universe, you are entering regions of stronger gravity and greater curvature of space. Indeed, the portions of your spacelike slice that cut through the event horizon must have considerable curvature. These highly curved central portions of your spacelike slice constitute the "throat" of the Einstein-Rosen bridge.

There is one final characteristic of the Penrose diagram in Figure 6-5 that should not be overlooked. In the previous chapter we learned that the directions of space and time are interchanged upon crossing an event horizon. This fact is clearly displayed in our Penrose diagram. Under normal circumstances, a line indicating the location of an object is oriented vertically in a spacetime diagram. For example, think about your own path in an ordinary spacetime diagram like Figure 4-4. If you move neither to the right nor to the left, you simply follow a vertical path as you get older and older. The line giving your location is vertical.

But inside the event horizon of a static black hole, the space and time directions are interchanged. Consequently, the lines giving the locations of the singularities must be horizontal! This is why an infalling astronaut is forced to smash into the singularity inside a static black hole. The horizontal singularity permanently cuts off any possibility of returning to flat spacetime.

It is intriguing to note that the same miserable fate does not necessarily befall an astronaut who plunges into an electrically charged black hole. An electrically charged black hole has *two* event horizons that separate the singularity from the outside universe. Therefore, in approaching the hole's center, an astronaut finds that there are *two* complete interchanges of spacelike and timelike directions. With two complete interchanges, the directions of space and time inside the inner event horizon are the *same* as they were far from the hole. Consequently, the singularity of electrically charged black holes must be oriented *vertically* in a Penrose diagram.

A Penrose diagram of an electrically charged black hole is shown in Figure 6-6. As we expected, the singularities are oriented vertically. And as usual, the event horizons are inclined at 45°.

The extraordinary direct result of vertically oriented singularities is that you *can* get from our universe through the charged wormhole to another universe. Of course you will encounter strong gravity, but not necessarily infinite curvature as you thread your way along an allowed path in spacetime inclined by less than 45° from the vertical.

Although gravity does not necessarily crush you to smithereens inside a Reissner-Nordstrøm hole, the powerful electric fields could easily be strong enough to tear apart every atom in your body. As negatively charged electrons are ripped away from their positively charged nuclei, an adventurous astronaut finds that his or her journey into a Reissner-Nordstrøm hole has indeed become exceedingly unpleasant! We shall therefore defer any discussion of trips to other universes to rotating black holes.

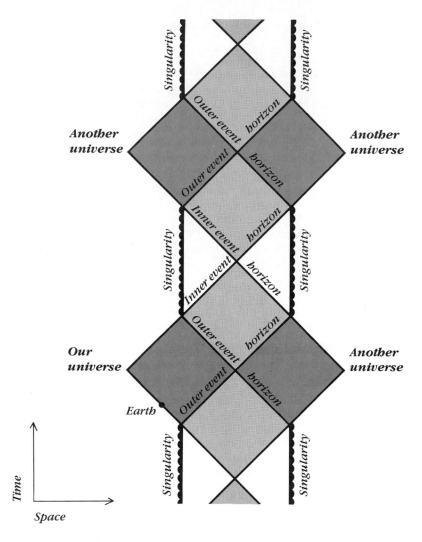

**Figure 6-6   The Spacetime Map of an Electrically
Charged Black Hole**

*This map shows regions of spacetime associated with an electrically
charged (Reissner-Nordstrøm) black hole. The diagram repeats over
and over into the past and into the future. The singularities are
oriented vertically (that is, in the timelike direction) because two
event horizons separate the singularities from the outside universes.*

A rotating black hole shares some important characteristics with a charged black hole. For example, both have two event horizons. Since a rotating black hole has two event horizons, its singularity must also be oriented vertically in a Penrose diagram. In fact, the Penrose diagram for a Kerr black hole (see Figure 6-7) is very similar to that of a Reissner-Nordstrøm hole. One important difference lies in the fact that a rotating black hole has a ring-shaped singularity instead of a point singularity the way static or charged holes have. You encounter infinite spacetime curvature inside a rotating hole only if you approach the ring singularity edge-on. At any other angle, you can travel through the center of the hole. In doing so, you go from ordinary space (where, for example, you could be located 10 meters from the singularity) into negative space (where, for example, you could be located − 10 meters from the singularity). This negative character of the space on the other side of the ring singularity means that gravity is repulsive. Plunging through the ring singularity, you find that you have entered an antigravity universe where you are vigorously pushed away from the hole!

Trip *A* in Figure 6-7 is one of these journeys into an antigravity universe. You leave the earth and plunge first through the outer event horizon and then through the inner event horizon. While inside the inner event horizon, you aim your spaceship toward the hole's center and pierce the ring singularity. Of course, you are careful to approach the hole's center at an oblique angle, thereby avoiding infinitely curved spacetime. Because you can either strike or miss infinitely curved spacetime, depending on the angle of approach, the singularities in the Penrose diagram of a rotating hole are usually indicated by dashed lines.

Trip *B* in Figure 6-7 simply takes you to another universe associated with the geometry of the rotating hole. You move inward through the outer event horizon and then inward through the inner event horizon. Since the singularity is oriented vertically, there is nothing to prevent you from continuing. You avoid the singularity completely and simply exit through the inner event horizon and then through the outer event horizon. Once outside the outer event horizon, you can journey farther and farther from the hole and explore the universe into which you have emerged.

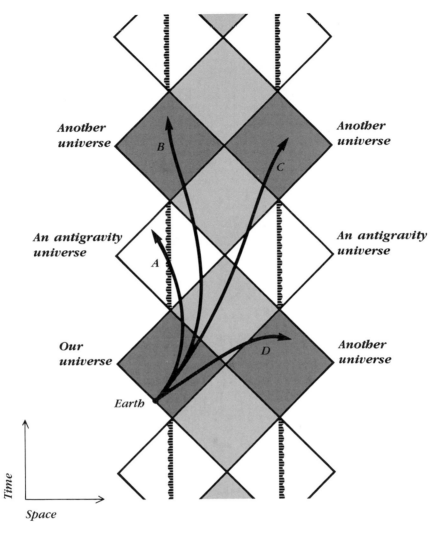

Figure 6-7   The Spacetime Map of a Rotating Black Hole
The spacetime maps of electrically charged and rotating (Kerr)
black holes are very similar because both types of holes have inner
and outer event horizons. The main difference involves the
character of the singularity. In a rotating black hole, it is possible
to pass through the ring singularity and enter an antigravity
universe, as shown in trip A. Trips B and C simply carry
adventurous astronauts to other ordinary universes. Trip D is
forbidden because it involves faster-than-light space travel.

Trip *C* is basically the same as trip *B* except that you come out of the hole into a different universe. Because of the similarities between Penrose diagrams of charged and rotating holes (compare Figures 6-6 and 6-7), journeys similar to trips *B* and *C* would also be possible through a charged hole. Of course, trip *D* is forbidden. Trip *D* is inclined by more than 45° from the vertical and would require a faster-than-light spaceship.

A fascinating aspect of these Penrose diagrams deals with the interpretation of the "other universes." One possibility is that they are indeed separate regions of space and time that are completely disconnected and totally unrelated to our universe. But an equally acceptable idea is that they are all our own universe! This is just like saying that the upper and lower sheets of an Einstein-Rosen bridge are really distant portions of our own universe connected by a wormhole. Perhaps, therefore, a rotating hole connects our universe with itself in a multitude of places! But remember, these would be different places in space and/or time. In other words, by emerging into one of these "other universes," you might actually be reentering our own universe in the same place but at a different time. *This is a time machine!* Theoretically, by plunging into a rotating wormhole and by carefully piloting your spacecraft, you could reemerge into our universe a billion years ago and visit the earth before the age of dinosaurs. Or you could reemerge a billion years in the future and meet the creatures that eventually evolved from the lower life forms that today we call human beings.

Is this really possible? Are some of these fantastical predictions really valid? The Penrose diagrams are, after all, a direct, logical consequence of our best theory of gravity: the general theory of relativity. But are we to believe all this?

There are a few problems. For example, if you could use a rotating wormhole as a time machine and come back a billion years ago, then you could also certainly arrange to come back to the earth an hour before you left. You could meet yourself and tell yourself what a fine trip you had! And then both of you could get on the rocketship and take the trip again! And again! And again!

Clearly, this would be a very strange state of affairs. We therefore wonder if perhaps we have overlooked some important factors.

Specifically, we notice that while traveling through a rotating wormhole (like trip *B* in Figure 6-7), it is possible to cruise very near infinitely curved spacetime without falling in. What is it like to stand near the singularity? What kind of processes occur near infinitely warped spacetime?

As we shall see in a later chapter, powerful gravitational fields do indeed have a profound effect on their surroundings. Many physicists believe that these effects choke off the wormhole, thereby preventing any fantastical trips. But we shall see that these same phenomena also cause small black holes to evaporate and explode.

7

The Search
for Black Holes

As you can perhaps imagine, finding black holes in space is a difficult business. You cannot simply go to a telescope and look up at the black sky honestly expecting to find a black hole. The fact that black holes do not emit light makes them extremely elusive.

Of course, all black holes are surrounded by intense gravitational fields. Consequently, it might be possible to discover a black hole by observing the effects of its strong gravity on its surroundings. For example, gases falling toward a black hole might be detectable. Unfortunately, this would amount to an *indirect* discovery, because the hole itself would never be observed. Astronomers must therefore be very careful to rule out alternative non-black-hole explanations of the phenomena they observe. These alternative explanations must be thoroughly examined and firmly rejected before we can feel confident that a black hole has been discovered.

One possible method for detecting black holes involves the gravitational deflection of light. We are familiar with the fact that light rays passing through a strong gravitational field are bent from their usual straight-line paths. We saw that this is a small effect in the case of light rays grazing the sun (recall Figure 4-6) because spacetime around the sun is only slightly curved. But for light rays passing near a black hole, this effect would be much more pronounced because of the substantial curvature of spacetime around the hole. In fact, as shown in Figure 7-1, there are at least two paths by which light rays from a distant star can graze a black hole and arrive at an observer here on the earth. An astronomer would therefore see two images of the distant star, one image on either side of the hole. This arrangement, whereby a black hole distorts the appearance of remote background stars or galaxies, is called a *gravitational lens*.

Many fascinating calculations have been done concerning gravitational lenses. Unfortunately, these calculations reveal that the alignment between the earth, the black hole, and the background star must be extremely precise in order for any noticeable effect to occur. Such perfect alignments are extremely rare, which probably explains why no one has yet observed a gravitational lens. Nevertheless, astronomers are armed with these predictions for future reference. For example, Figure 7-2 shows how the image of a remote galaxy would be distorted. Two lenticular images would be seen, one on either side

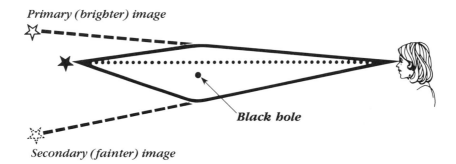

Primary (brighter) image

Black hole

Secondary (fainter) image

Figure 7-1 A Gravitational Lens
Light rays from a very distant star are deflected by a black hole's gravity on their way to the earth. An astronomer sees two images of the background star. The brighter image lies closer to the direct straight-line path between the star and the earth.

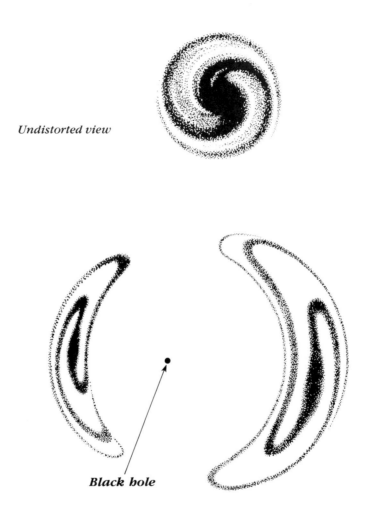

Undistorted view

Black hole

Figure 7-2   A Galaxy Seen Through a Gravitational Lens
The upper drawing shows the undistorted view of a remote spiral galaxy. If a black hole were located between us and the galaxy, we would see two lenticular images of the background galaxy. (Adapted from R. C. Wayte.)

of the hole. Of course, the black hole would have to be located nearly on a perfectly straight line between us and the background galaxy.

Perhaps a more fruitful attack on the problem of finding black holes would involve sucking up gas and dust rather than bending light rays. Perhaps gases plunging toward a black hole might emit radiation that could be detected from afar.

Individual, isolated black holes around our Galaxy would be nearly impossible to detect. Although a black hole can be considered a "cosmic vacuum cleaner" that sucks up everything in its path, interstellar space is almost perfectly empty. The atoms of interstellar gas are so widely separated that an isolated black hole could not swallow this gas fast enough to produce detectable radiation. Instead, we must hunt for black holes that are located near ample supplies of gas. Double star systems, therefore, are ideal candidates.

There are many examples of double star systems in which one star is rather ordinary but its companion has long ago lived out its life and become a stellar corpse. For example, in Chapter 2 we saw that novas occur in double star systems that contain a white dwarf. Gases from the ordinary companion star pile up on the white dwarf's surface. When these gases finally ignite, the white dwarf blazes forth with unprecedented brilliance. Prior to the outburst, astronomers usually had no idea that there was a white dwarf in that part of the sky.

In the same way, bursters are double star systems that contain a neutron star. As we saw in Chapter 3, powerful bursts of X rays are produced when gases from the ordinary companion star pile up on the neutron star's surface. These blasts of X rays announce the existence of neutron stars that otherwise would have remained completely undetectable.

On Saturday, December 12, 1970, the United States launched a small satellite into orbit from a platform in the Indian Ocean, just off the coast of Kenya. The date was the seventh anniversary of Kenyan independence, and the satellite was christened *Uhuru,* which means "freedom" in Swahili. Uhuru was the first in a series of astronomical satellites designed to detect X rays from objects in space.

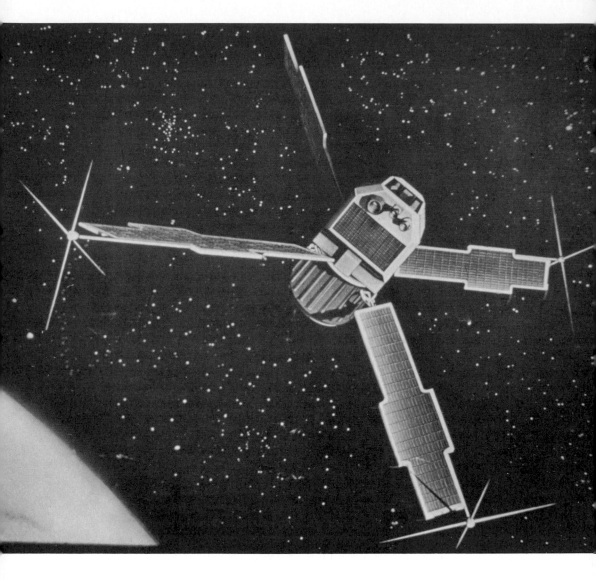

Figure 7-3 Uburu
This small satellite was designed to detect X rays from stars and galaxies. The satellite operated for 2 ½ years and observed 339 separate X-ray sources around the sky. (NASA.)

Earth-bound astronomers are condemned to view the heavens through an ocean of air. At optical wavelengths our atmosphere degrades the view only slightly. Images of stars and galaxies twinkle and shimmer because of turbulence and currents in layers of nitrogen and oxygen that we breathe. Although this is a relatively minor annoyance at optical wavelengths, our atmosphere completely prohibits any earth-based observations at X-ray wavelengths. The air is opaque to X rays, and consequently astronomers must make their observations above the earth's atmosphere. Uhuru was destined to give us our first complete look at the X-ray sky.

Uhuru consisted of two X-ray telescopes positioned back to back. As the satellite slowly rotated, the telescopes scanned the skies. When a source of X rays came into view, signals were transmitted to astronomers on the ground. By knowing the orientation of the satellite in space, the astronomers could then deduce the position of the X-ray source. Before its battery and transmitter failed in the spring of 1973, Uhuru had succeeded in locating 339 sources of X rays across the sky.

The X-ray sources discovered by Uhuru are shown on a map of the sky in Figure 7-4. For comparison, a map of the visible sky is also included in this illustration. Both pictures are drawn so that the Milky Way extends horizontally across the middle of each map. It is therefore immediately clear that a large fraction of the sources discovered by Uhuru are located in the plane of the Milky Way along with most of the stars in our Galaxy. The majority of these Milky Way sources are probably stellar corpses such as white dwarfs and neutron stars. In contrast, however, most of the sources positioned above and below the Milky Way are identified with extremely remote objects such as exploding galaxies and quasars.

Less than a year after Uhuru's launch, much attention became focused on one exceptionally bright X-ray source, called Cygnus X-1. This source (the brightest X-ray object in the constellation of Cygnus) was particularly intriguing because it flickers very rapidly, more rapidly than once every thousandth of a second.

A basic premise at the foundation of all modern science is that nothing can travel faster than the speed of light. This means that

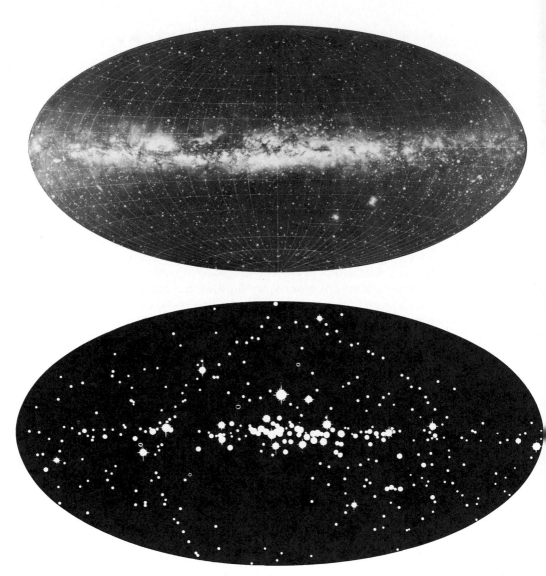

Figure 7-4 The Visible Sky and the X-Ray Sky
These matching illustrations contrast the appearance of the visible sky (upper) and the X-ray sky (lower). Every visible star appears in the upper map. All the X-ray sources discovered by Uhuru are shown in the lower map. (Lund Observatory.)

an object cannot flicker faster than the light-travel time across the object. For example, if an object is 1 light year in size, it cannot change its brightness faster than once a year. Using this line of argument, astronomers can place firm limits on the maximum possible sizes of objects they observe in the sky. Specifically, the rapid flickering of Cygnus X-1 means that the source of X rays must be smaller than one thousandth of a light second in diameter. Cygnus X-1 must be smaller than 300 kilometers across.

The realization that one of the most powerful sources of X rays in the sky is less than 300 kilometers in diameter inspired teams of astronomers to try to pin down the exact location of Cygnus X-1. By the end of 1971, it was clear that Cygnus X-1 coincides with the bright, hot, bluish star called HDE 226868, shown in Figure 7-5. But ordinary stars do not emit any appreciable amount of X rays. In addition, hot, bright stars like HDE 226868 are considerably larger than the sun. Consequently, the visible star identified in Figure 7-5 cannot itself be the source of X rays. Astronomers therefore began suspecting that the X rays were actually coming from an unseen companion star in orbit about HDE 226868.

Detailed analysis of the light from HDE 226868 soon confirmed these suspicions. It is indeed a double star system. The two stars are actually quite close together, separated by a mere 30 million kilometers (that is about one-fifth of the distance from the earth to the sun). At such close quarters, the two stars rapidly revolve about their common center, taking only 5.6 days to complete an orbit.

Double stars have traditionally played an important role in astronomy. By carefully observing the orbits of two stars about their common center, it is possible to deduce the masses of the stars. Unfortunately, only one star is visible in the Cygnus X-1 system, and therefore astronomers are not able to determine the stellar masses as precisely as they would like. Nevertheless, it is clear that the visible star HDE 226868 has a mass of roughly 20 solar masses. That is rather typical for hot, bright, bluish stars. And the unseen, X-ray-emitting companion has a mass of roughly 10 solar masses. There is only one kind of object that contains 10 solar masses compressed into a volume smaller than 300 kilometers across. Cygnus X-1 must be a black hole.

Figure 7-5   HDE 226868
*Teamwork by radio astronomers and X-ray astronomers in 1971
gave a very accurate determination of the position of Cygnus X-1: it
has the same location as the star HDE 226868 (indicated by the
arrow). Observations in 1972 showed that HDE 226868 is one
member of a double star system. The invisible companion is Cygnus
X-1. (Courtesy of J. Kristian, Hale Observatories.)*

If Cygnus X-1 is a black hole, where do the X rays come from? They certainly do not come from the hole itself, because nothing can escape from a black hole.

The past two decades of space exploration have been filled with a multitude of surprises and unexpected discoveries. We have found craters on Mercury, volcanoes on Mars, and a ring around Jupiter. And during the silent journeys across interplanetary space, sensors on the spacecrafts report a continuous streaming of particles from the sun. This constant flow of particles (mostly electrons and protons) from the sun's outermost layers is called the *solar wind*. Because of the solar wind, the sun is gradually shedding some of its gases, albeit at an almost insignificant rate.

Since the sun is an ordinary, garden-variety star, astronomers have come to realize that all stars probably possess *stellar winds*. All stars are gradually shedding their outermost layers in wispy streams of protons and electrons. Specifically, HDE 226868 should possess a stellar wind. As particles of this wind pass near the black hole, they are captured into orbit about the hole. We therefore expect the hole to be surrounded by a large disk of gases accumulated over the years, like a giant version of the rings around Saturn, as shown in Figure 7-6.

In thinking about the structure of this disk of gases, we realize that the inner portions of the disk should be revolving about the hole much more rapidly than the outer portions. This difference in speed occurs for exactly the same reason that Mercury revolves about the sun much more rapidly than Pluto does. Furthermore, the rapidly moving gases nearer the hole should be constantly rubbing against the slower-moving gas farther from the hole. This friction has two effects. First, it causes the gases to spiral in toward the hole. But in addition, these inward-spiraling gases are heated by this friction to higher and higher temperatures as they get closer and closer to the hole. In fact, detailed calculations reveal that these gases reach temperatures of nearly 2 million degrees at the innermost regions of the disk, just above the black hole. Anything at such enormous temperatures shines brightly in X rays. The X rays detected by Uhuru are therefore produced by extremely hot gases just before they take the final plunge down the black hole.

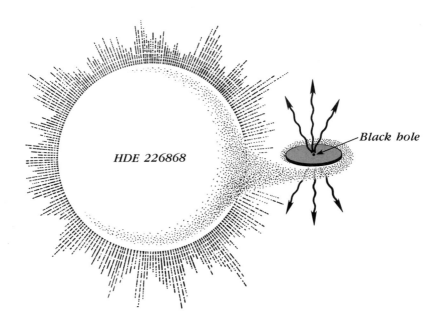

Figure* 7-6   *The Cygnus X-1 Black Hole
*The stellar wind from HDE 226868 pours matter onto a huge
disk around its black hole companion. The infalling gases are heated
to enormous temperatures as they spiral toward the black hole. At
the inner edge of the disk, just above the black hole, the gases are so
hot that they emit vast quantities of X rays.*

In the years following Uhuru, a series of X-ray-detecting satellites have been launched into space. Each of these satellites is an improvement over its predecessor, and we are constantly treated to better and better views of the X-ray sky. For example, one of the finest X-ray satellites, initially called the second High Energy Astronomical Observatory, or HEAO 2, was launched into a 537-kilometer-high equatorial orbit on November 13, 1978. In honor of the one hundredth anniversary of Albert Einstein's birth on March 14, 1879, this huge satellite has been christened the *Einstein Observatory.*

The Einstein Observatory and its predecessors have uncovered a number of X-ray sources that bear striking resemblances to Cygnus X-1. Several of these sources are identified on the star charts in Figure 7-7. For example, V861 Scorpii is the visible star in a binary system that contains an X-ray companion. The orbital period is 8.4 days, and the mass of the unseen companion is computed between 5 and 12 solar masses. This means that the X-ray companion of V861 Scorpii is far too massive to be either a white dwarf or a neutron star. The only remaining choice is a black hole.

Ideally, astronomers would like to find black hole candidates in double star systems where the ordinary star is easily observed through ordinary optical telescopes. This would allow us to measure the stellar orbits and compute the mass of the unseen X-ray object. If the answer is clearly greater than the upper limit for either a white dwarf or a neutron star, then we can feel confident that another black hole has been discovered. Unfortunately, nature is often unaccommodating, and the ordinary star is too dim to be seen. Nevertheless, it is still possible to ferret out good black hole candidates.

Several X-ray sources exhibit sporadic short-term variability. They flicker very rapidly. This rapid X-ray flickering was precisely the characteristic that first drew attention to Cygnus X-1 in the early 1970s. It means that the source of X rays is very small, far too small to be a white dwarf. In addition, if the source does not exhibit regular pulses of radiation, then it is probably not a neutron star.

Circinus X-1 and GX 339-4 are fine examples of X-ray sources that exhibit sporadic flickering just like Cygnus X-1. Astronomers speculate that each of these sources is a member of a double star

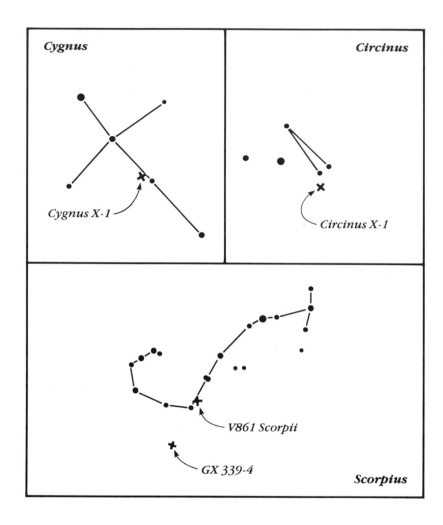

Figure 7-7   Four Black Hole Candidates
These star maps give the locations of four X-ray sources that may in fact be black holes. The case for Cygnus X-1 is the strongest.

system that contains a black hole surrounded by a disk of captured gases. As in the case of Cygnus X-1, the rapid flickering is caused by hot spots on the inner edge of the disk. As these random hot spots whiz around the hole, they are occasionally exposed to our view, and their X rays are gravitationally focused by the hole.

This shortcut method of looking for rapid X-ray flickering allows us to pick out black hole candidates from among the dozens of X-ray sources that are constantly being discovered across the sky. And we may soon find that black holes are far more plentiful than anyone had ever imagined.

8

Exploring the Cosmos

The star-filled night sky is surely among the most awe-inspiring phenomena in the world around us. For this reason alone, the study of astronomy was destined to predate the dawn of recorded history.

Although astronomy is one of the most ancient endeavors of the human mind, it is also an exciting, fast-paced area of modern scientific research. Indeed, what we have learned about the universe in the past few decades is more than all the knowledge accumulated over the preceding eons of stargazing. And our ideas about the cosmos have undergone radical and sometimes abrupt changes in direct response to these twentieth-century revelations. For example, up until the mid-1920s, we did not really know about the existence of galaxies. In spite of some correct speculation, the pre-1920 data were inconclusive, and many professional astronomers refused to accept the idea that galaxies are scattered across the cosmos.

We now know that galaxies are the largest individual objects in the universe. A typical galaxy measures 100,000 light years in diameter and contains roughly 200 billion stars.

There are basically four types of galaxies in the universe. First, there are the *spiral galaxies* with graceful spiral arms outlining their familiar pinwheel shapes. A fine example is shown in Figure 8-1. As this photograph strongly suggests, the stars in a galaxy revolve about the galaxy's nucleus just as the planets orbit the sun. But galaxies are so huge that it usually takes hundreds of millions of years for a star to complete one circuit about its galaxy's center.

We are living inside one of these spiral galaxies. As we learned in Chapter 1, we are located about 30,000 light years from our Galaxy's nucleus, nestled in between two spiral arms. In fact, the Milky Way that you see arching overhead at night is our inside view of these two neighboring spiral arms. Of course, we and all the nearby stars are lumbering along together about the galactic center. Our speed is 300 kilometers per second (which is about 670,000 miles per hour), and it takes us 250 million years to complete one full orbit. Since the earth is about 5 billion years old, our planet has been around the Galaxy twenty times.

Although they are in a distinct minority, there is another kind of spiral galaxy. In this second type of galaxy, the spiral arms originate at the ends of a bar of stars and gas that extends across the galaxy's

Figure 8-1   A Spiral Galaxy
*This galaxy (called M101 or NGC 5457) is located in the
constellation of Ursa Major. The graceful, arching spiral arms are
outlined by numerous vast clouds of glowing gases. (Kitt Peak
National Observatory.)*

nucleus. Galaxies with this characteristic appearance are appropriately called *barred spiral galaxies.* An excellent example is shown in Figure 8-2.

Spiral and barred spiral galaxies together account for only a third of all the galaxies in the universe. *Elliptical galaxies* are clearly in the majority. Unlike their spectacular spiral cousins, however, elliptical galaxies do not have a particularly dramatic appearance. Galaxies of this third type do not possess spiral arms or dust lanes or emission nebulas among their stars. In fact, these galaxies usually look like uninteresting blobs. As their name suggests, they have elliptical shapes that can range from completely circular to highly flattened. Figure 8-3 shows an elliptical galaxy that is nearly circular.

Finally, about 10 percent of the galaxies in the universe do not fall into any of these three categories. These oddballs often have a very distorted appearance that earns them the separate classification of *irregular galaxies.* These deformed cosmic freaks often owe their twisted shapes to some violent processes occurring deep in their interiors. Because of their pathology, they are among the most fascinating and fantastical objects in the universe. Many of them will be discussed in detail in the next chapter.

Galaxies are not scattered randomly across space. Instead, they are grouped together in huge aggregates called *clusters.* Examples are shown in Figures 8-4 and 8-5. Some clusters have a distinctly spherical shape. These are called *regular clusters* and usually have a high abundance of elliptical galaxies. In contrast, *irregular clusters* are sprawling assemblages, consisting of a more democratic mixture of all types of galaxies.

We live in an irregular cluster of galaxies affectionately known as the *Local Group.* The Local Group is called a "poor" cluster because it contains only two dozen galaxies. The largest member is the Andromeda Galaxy, shown in Figure 8-6. As far as spiral galaxies go, it is a whopper. Its mass is estimated at 340 million solar masses.

The Andromeda Galaxy is positioned on one side of the Local Group. We are located on the other side, about 2¼ million light years away. There is one other spiral galaxy and four irregular galaxies in our Local Group. But by far, the majority are small and dwarf elliptical galaxies. An example is shown in Figure 8-7.

Figure 8-2   A Barred Spiral Galaxy
*This galaxy (called NGC 1300) is located in the constellation of
Eridanus. Notice that the spiral arms originate at the ends of a bar
of stars and glowing gas that extends through the galaxy's nucleus.
(Hale Observatories.)*

135

Figure 8-3   An Elliptical Galaxy
This galaxy (called M49 or NGC 4472) is located in the constellation of Virgo. Elliptical galaxies have almost no dust or gas between the stars. Although this galaxy is very circular, some elliptical galaxies have a flattened, cigar-shaped appearance. (Kitt Peak National Observatory.)

Figure 8-4   A Regular Cluster of Galaxies
This rich cluster of galaxies is located in the constellation of
Corona Borealis. Like all regular clusters, this cluster has an overall
spherical shape and possesses a high percentage of elliptical
galaxies. (Hale Observatories.)

Figure 8-5   An Irregular Cluster of Galaxies
*This cluster of galaxies is located in the constellation of Hercules.
Like all irregular clusters, this cluster has a sprawling irregular
shape. Notice the higher percentage of spiral galaxies compared
with the regular cluster in Figure 8-4. (Hale Observatories.)*

Figure 8-6   The Andromeda Galaxy
*This nearby galaxy (also called M31 or NGC 224) is located in the
constellation of Andromeda, at a distance of only 2 ¼ million light
years. It is the largest galaxy in the Local Group. (Lick
Observatory.)*

139

Figure 8-7   A Small, Nearby Elliptical Galaxy
*This galaxy (called NGC 147) is so nearby that individual stars are
clearly seen. It is located in the constellation of Cassiopeia at a
distance of only 1.9 million light years, slightly nearer to us than
the Andromeda Galaxy is. (Hale Observatories.)*

If you could journey far into space, far beyond the stars, to a distance of several million light years, you could actually see the entire Local Group spread out before you. Each galaxy would be a hazy patch of light against the blackened, starless sky. One hazy patch would be our Galaxy; another surely would be Andromeda. And maybe you might see a few others, but all the dwarf elliptical galaxies probably would go unnoticed. These galactic midgets contain so few stars that you can see directly through their centers with no trouble at all. Because of their low luminosity, thousands upon thousands of dwarf ellipticals have certainly escaped detection in even the nearest clusters that have been so extensively studied over the past couple of decades.

The first major twentieth-century revolution in our ideas of the cosmos occurred when we realized that galaxies exist. It was suddenly clear that the universe is far bigger than anyone had dared to imagine. Indeed, it is vast on an inconceivable scale. Humanity shall never be able to return to the quaint, limited, circumscribed cosmologies that dominated ancient myths and religions.

The second revolution came when we discovered how the galaxies are moving. We then realized that our inconceivably vast universe is actually expanding.

The motion of a star or galaxy can be measured by spectroscopically analyzing its light. This is accomplished by passing the starlight through a prism, which breaks up the light into the colors of the rainbow, as shown in Figure 8-9. Detailed examination of this rainbow (technically called a *spectrum*) often reveals several dark *spectral lines* interspersed among the colors. These spectral lines are caused by the chemicals of which the star or galaxy is made. Each chemical element produces its own characteristic set of spectral lines. This is how we know what the stars are made of. For example, if we see spectral lines of silicon in the spectrum of a certain star, then we know that silicon must be present in that star's gases.

Light is a phenomenon that involves waves, analogous to water waves that ripple across a lake or the ocean. But instead of water molecules bobbing up and down, a light ray consists of vibrating electric and magnetic fields. Furthermore, the color of light that

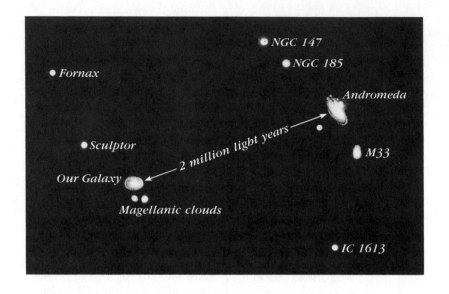

Figure 8-8   The Local Group
This drawing is an artistic representation of what a remote alien astronomer looking toward our Galaxy would see. All the dwarf ellipticals would probably escape detection.

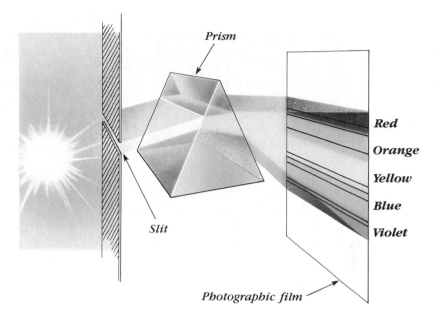

Prism

Red

Orange

Yellow

Blue

Violet

Slit

Photographic film

Figure 8-9   *A Spectrum and Spectral Lines*
When light from a star or galaxy is passed through a prism, the light is dispersed into the colors of the rainbow. This spectrum often contains dark spectral lines that were produced by atoms in the star or galaxy. The spectrograph that an astronomer uses at a telescope also has several lenses to magnify and focus the spectrum onto a photographic plate.

you see depends entirely on the *wavelength* of that light. As shown in Figure 8-10, the wavelength is simply the distance between two adjacent crests of the wave. Of all the colors of the rainbow, violet light has the shortest wavelength, only 0.0004 millimeter. And at the other end of the spectrum, red light has the longest wavelength, nearly 0.0008 millimeter. Intermediate colors of the spectrum have intermediate wavelengths.

While the patterns of spectral lines divulge the chemical composition of a source of light, the exact positions of these spectral lines reveal how the source of light is moving. To see why this is so, imagine a source of light coming toward you. As illustrated in Figure 8-11, the waves emitted from the moving source are crowded together in front of the source. Consequently, when you examine the spectrum of this approaching light, you find that all the spectral lines are located at shorter wavelengths than usual.

A displacement of spectral lines toward shorter-than-usual wavelengths is called a *blueshift* simply because blue light has one of the shortest wavelengths of all the colors of the rainbow. Furthermore, the amount of the shift is directly related to the speed of the source of light. Of course, if the source of light is not moving, the spectral lines are positioned exactly where they should be among the colors of the rainbow. (The proper positions of spectral lines have been determined in laboratory experiments and are listed in reference books to which astronomers often refer.) But if the source of light is approaching you, all the spectral lines are shifted toward the blue end of the spectrum. The higher the speed, the greater the shift.

In exactly the same fashion, if a source of light is moving away from you, all its waves are stretched out behind the receding source (see Figure 8-11). This means that all the spectral lines in the source's spectrum are displaced toward longer-than-usual wavelengths. This is appropriately called a *redshift* because red light has the longest wavelength of all the colors of the rainbow. And once again, the higher the speed, the greater the shift.

In the late 1920s, a young astronomer at Mount Wilson Observatory drew attention to a systematic effect involving galaxies and their spectra. Edwin Hubble found that nearby galaxies have their spectral lines almost exactly where they should be among the colors

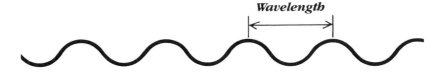

Wavelength

Figure 8-10   Waves and Wavelength
Light is a wave phenomenon. The wavelength of a particular
color of light is the distance between adjacent crests (or troughs) in
the wave. Wavelengths of visible light range from 0.0004 millimeter
(for violet light) to nearly 0.0008 millimeter (for red light).

of the rainbow. But as he turned his attention to more remote
galaxies, he found a predominance of redshifts. In fact, the more distant the galaxy, the higher its redshift.

This phenomenon is shown in Figure 8-12. The photographs
of five elliptical galaxies appear on the left of the illustration. The
spectrum of each galaxy is shown to the right. In each case, the
galaxy's spectrum is the hazy horizontal band. The bright vertical
lines above and below each band are the "comparison spectrum" that
the astronomer artificially exposes onto each spectroscopic plate. The
standard lines of the comparison spectrum serve as handy reference
markers from which the astronomer can measure the amount of displacement of the galaxy's spectral lines.

Each of the five spectra in Figure 8-12 shows two spectral
lines. They are the "H and K lines" of calcium. The proper position
for these two lines (as determined in laboratory experiments with
calcium) is among the blue colors of the rainbow, at the short-
wavelength side of the spectrum.

In the case of the nearest galaxy (in Virgo), the H and K lines
are almost exactly where they belong, among the blue colors of the
rainbow at the left side of the spectrum. But notice that for more

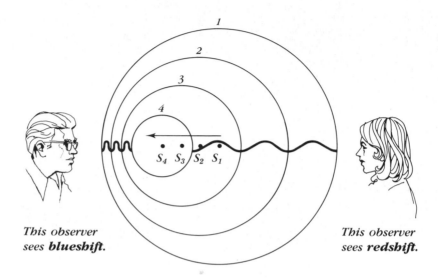

This observer sees **blueshift**.

This observer sees **redshift**.

Figure 8-11 The Doppler Effect
Radiation from an approaching source of light is compressed to shorter wavelengths than usual. Radiation from a receding source of light is stretched to longer wavelengths than usual. The amount of wavelength shift depends on the speed between the source and the observer. The higher the speed, the greater the shift.

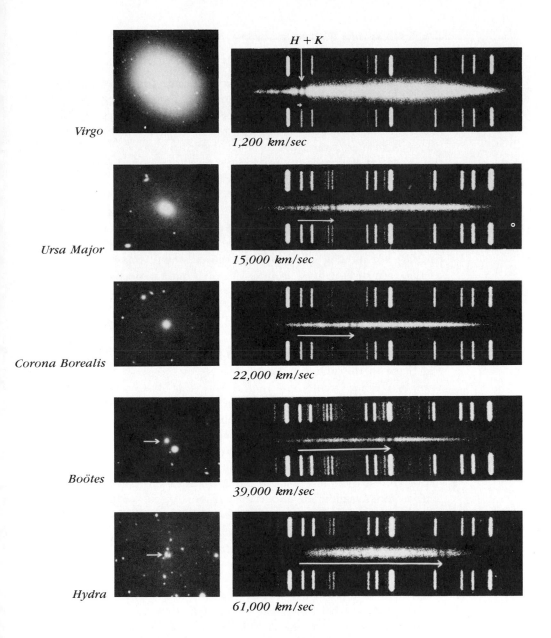

H + K

Virgo

1,200 km/sec

Ursa Major

15,000 km/sec

Corona Borealis

22,000 km/sec

Boötes

39,000 km/sec

Hydra

61,000 km/sec

Figure 8-12   Galaxies and Their Spectra
*Five galaxies (labeled according to the constellation in which they
are located) are shown along with their spectra. The nearest
galaxies have the smallest redshifts, while the more distant galaxies
have larger redshifts. (Hale Observatories.)*

distant galaxies, the H and K lines are shifted to longer wavelengths toward red colors of the rainbow on the right side of the spectra. In fact, for the most distant galaxy in this illustration (in Hydra), the H and K lines are displaced all the way across the spectrum. Clearly this galaxy is rushing away from us at a substantial velocity. In this case, the extent of the redshift corresponds to a speed of 61,000 kilometers per second (136 million miles per hour).

By the early 1930s, Edwin Hubble had established the fact that there is a direct linear relationship between the distances to galaxies and the size of the redshifts of the galaxies. This relationship is called the *Hubble law* and is shown graphically in Figure 8-13. Nearby galaxies are moving away from us rather slowly, but more distant galaxies are rushing away from us at much higher speeds.

This phenomenon exhibited in the Hubble law is not noticeable over small distances. You have to look far beyond the confines of the Local Group to detect the cosmological redshifts of galaxies. You have to gaze across distances of 100 million light years or more.

Of course, we are not located in a special galaxy or in a special place in the universe. Such ideas went out the window many years ago. Consequently, any alien astronomer in any galaxy will see all the other distant clusters of galaxies rushing away from him or her, just as we see all the remote clusters of galaxies rushing away from us. It is therefore clear that the true meaning of the Hubble law is that *we live in an expanding universe!*

The Hubble law can justifiably be considered the single most important astronomical discovery of the century. It is a fundamental and profound statement about the structure and evolution of the universe as a whole. Widely separated clusters of galaxies are getting farther and farther apart, thereby revealing the expansion of the universe on an inconceivably colossal scale.

Because the clusters of galaxies are getting farther and farther apart, there must have been a time in the remote past when they were piled on top of each other. Of course, individual galaxies probably did not exist back then. But by knowing how fast the universe is expanding (as given by the Hubble law in Figure 8-13), we can extrapolate back to a moment in time when the universe was in an infinitely dense state. This occurred 20 billion years ago. And at that

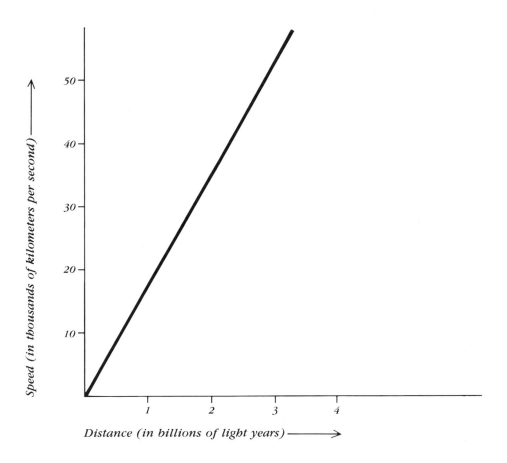

Figure 8-13   *The Hubble Law*
*The Hubble law is the direct proportionality between the speeds
and distances of galaxies. Nearby galaxies are moving away from
us slowly. More distant galaxies are rushing away from us with
correspondingly higher speeds.*

Figure 8-14   The Holmdel Horn Antenna
*This antenna was designed for relaying messages between the earth
and communication satellites. It was sensitive enough, however,
to pick up microwaves coming from all parts of the sky. These
microwaves are the echo of the primordial cosmic fireball. (Bell
Telephone Laboratories.)*

extraordinary moment an explosion must have occurred throughout all space to start the expansion that we observe today. This cosmic event is called the *Big Bang.*

In the early 1960s, physicists realized that the universe must have been extremely hot immediately after its creation. All space must have been filled with an intense radiation field at an enormous temperature. This was the *primordial fireball* that accompanied the birth of the universe. For example, 1 second after the Big Bang the temperature of the fireball must have been 10 billion degrees. But as the universe expanded, the fireball cooled as its radiations became more and more diluted over an ever-increasing volume of space.

In 1965, two physicists at Bell Telephone Laboratories in Holmdel, New Jersey, were working on a sensitive new antenna for use with communication satellites. This antenna is shown in Figure 8-14. Their careful work soon revealed that the entire sky is filled with microwaves. Although the intensity is very low, microwaves are coming to us from every part of the sky. This *cosmic microwave background* constitutes a field of radiation at a temperature of only 3 degrees above absolute zero.

The discovery of the microwave background is a powerful confirmation of our ideas about the Big Bang and the creation of the universe. The universe has been expanding for 20 billion years. And the primordial fireball, once so inconceivably hot, has cooled down to only 3 degrees above absolute zero. The eerie hiss of the cosmic static at microwave frequencies is the most ancient radiation our instruments shall ever detect. We shall never see anything beyond the microwave background. We have heard the echo of cosmic creation.

9

Quasars and Supermassive Black Holes

In 1931, a young engineer with Bell Telephone Laboratories was experimenting with antennas for long-range radio communication. During the course of his work, Karl Jansky discovered that these antennas were picking up radio static from an object in space. This was a momentous event in the history of astronomy. It was the first time that anyone had detected any *nonvisible* radiation from space. Prior to Jansky's discovery, everything we knew about the universe was based entirely on optical observations using visible light. But visible light is only a small fraction of all the types of radiation that could be coming to us from space. Jansky's discovery therefore opened up the possibility of detecting radio waves from objects across the universe. No longer would we be forced to rely on optical observations alone. We would be able to see the radio sky and view the universe at wavelengths to which our ordinary eyes are blind.

During the 1950s, radio telescopes began springing up in Australia, England, the Netherlands, and the United States. These devices (see Figure 9-1) collect and focus radio waves in much the same way that ordinary telescopes collect and focus ordinary light. With these radio telescopes, humanity began to get its first view of the invisible radio sky.

As radio astronomers began exploring the heavens, they found that many familiar objects are sources of radio waves. Most galaxies emit radio radiation. So do many nebulas. And there are also radio waves coming from locations where no visible objects have ever been identified, in spite of extensive searches with our largest optical telescopes. Contrasting views of the visible and radio sky are shown in Figure 9-2.

By 1960, radio astronomers began focusing their attention on a few radio sources that seemed to be a bit unusual. Most radio-emitting objects, such as galaxies and certain nebulas, are spread out over small areas around the sky. This made good sense. Radio waves are produced when electrons move through a magnetic field. Since the magnetic fields across galaxies are quite weak, and since there are very few electrons per cubic centimeter in space, a large volume is needed to produce a detectable radio signal. At stellar or galactic distances, this volume translates into a small patch of the sky. Indeed, virtually all of the radio sources discovered and identified prior to

1960 seemed to be smeared out over small areas. Astronomers were therefore mystified to find a few radio sources that seemed to be pinpoints of radio waves.

The first of these pinpoint radio sources to be discovered was 3C48 (so named because it is the 48th object listed in the *Third Cambridge Catalogue* of radio sources). Astronomers had little difficulty establishing that the radio waves were coming from the starlike object shown in Figure 9-3. But that's impossible! Ordinary stars simply do not produce enough radio waves to be detected by the radio telescopes of the late 1950s and early 1960s. And to make matters worse, no one could identify the spectral lines in the spectrum of 3C48. Although some bizarre (and totally incorrect) theories were proposed, no one could make any sense of 3C48 or its spectrum.

Within two years, a second mysterious point source of radio waves had been identified. Once again, the radio waves were coming from a starlike object. As shown in Figure 9-4, this object (called 3C273 because it is the 273rd radio source listed in the *Third Cambridge Catalogue*) has a strange luminous "jet" protruding to one side. And once again, no one could identify any of the spectral lines of 3C273.

In retrospect, we now realize that the main stumbling block in identifying the spectral lines of 3C48 and 3C273 was that everyone believed these objects to be nearby stars. Such stars would, of course, be moving along with the sun around our Galaxy. And thus, no one ever considered the possibility of substantial redshifts (or blueshifts) of the spectral lines to unfamiliar portions of the spectrum.

The breakthrough came in 1963. Maarten Schmidt at Caltech noticed that four prominent lines in the spectrum of 3C273 had exactly the same pattern and spacing as four of the most familiar spectral lines of hydrogen. But they were in absolutely the wrong place among the colors of the rainbow. The entire pattern was shifted by a colossal 16 percent toward the red end of the spectrum. This would mean that 3C273 is rushing away from us at the incredible speed of 15 percent of the speed of light.

Schmidt promptly found that the remaining lines of the spectrum of 3C273 are well-known spectral lines subjected to the same colossal redshift as the hydrogen lines. And within a few days, all of

Figure 9-1   A Radio Telescope
The dish of a radio telescope collects and focuses radio waves from space. Electronic equipment is used to amplify and record the signal. This radio telescope is at the National Radio Astronomy Observatory in West Virginia. The diameter of the dish is 140 feet. (N.R.A.O.)

156

Figure 9-2   The Visible Sky and the Radio Sky
The upper drawing shows the entire visible sky. The lower drawing shows the entire radio sky. Both illustrations have the same scale and orientation. Both are centered about the Milky Way. (Lund Observatory; Lois Cohen, Griffith Observatory.)

157

Figure 9-3 The Quasar 3C48
For several years astronomers believed erroneously that this object was simply a nearby peculiar star that just happens to emit radio waves. Actually, the redshift of this starlike object is so great that, according to the Hubble law, it must be roughly 5 billion light years away. (Hale Observatories.)

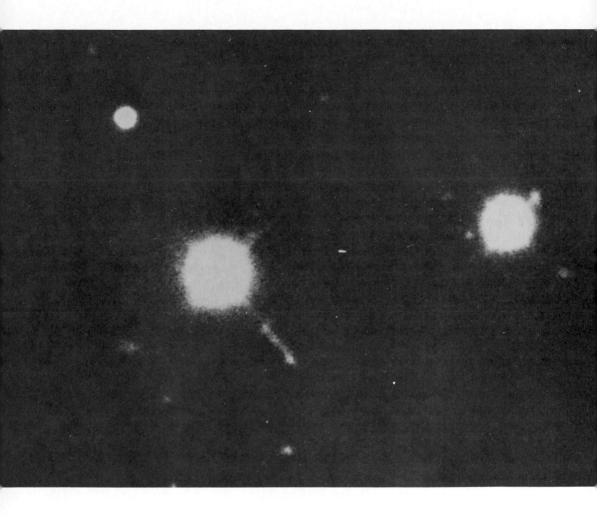

Figure 9-4   The Quasar 3C273
This highly enlarged view shows the starlike object associated with the radio source 3C273. Notice the luminous jet to one side of the "star." By 1963, astronomers discovered that the redshift of this "star" is so great that, according to the Hubble law, the star must be nearly 3 billion light years from the earth. (Kitt Peak National Observatory.)

159

the mysterious lines in the spectrum of 3C48 were also identified as familiar lines that had also suffered a huge redshift (this time, by 37 percent). This means that 3C48 must be rushing away from us at nearly one-third of the speed of light.

Obviously these objects are not stars. It is flatly impossible for ordinary stars to have such huge redshifts. But they certainly look like stars. They were christened *quasi-stellar objects,* a term that was soon shortened to *quasars.*

Since those early days, hundreds of quasars have been discovered all across the sky. All of them look like stars and all of them have huge redshifts. According to the Hubble law (recall Figure 8-13), quasars must therefore be very far away. Their high redshifts mean that they are rushing away from us very rapidly, which in turn (according to the Hubble law) places them at enormous distances from the earth. For example, 3C273 is 3 billion light years away, and the distance to 3C48 must be 5 billion light years.

One of the highest redshift quasars ever discovered, OH471, shown in Figure 9-5, has a redshift corresponding to a speed of 90 percent of the speed of light. According to the Hubble law, this huge recessional velocity gives the incredible distance of 16 billion light years from earth. Aside from the microwave background itself, OH471 is one of the most remote objects ever observed.

Quasars are the most distant individual objects that astronomers have ever observed. Specifically, we have never seen any galaxies with redshifts nearly as high as the record-breaking redshifts possessed by quasars. The reason is simple: ordinary galaxies are just too faint to be seen at distances of billions upon billions of light years. For example, if there are any galaxies out around OH471, they are far too faint to be detected even with our most powerful telescopes. Yet we have no trouble photographing quasars at these distances. Obviously, therefore, *quasars are the most luminous objects in the universe.* Indeed, a typical quasar shines with the brightness of a hundred ordinary galaxies. This is why we can see these objects at distances where ordinary galaxies are completely undetectable.

Incredible cosmic distance is certainly one of the intriguing characteristics of quasars. But the mystery deepened as soon as as-

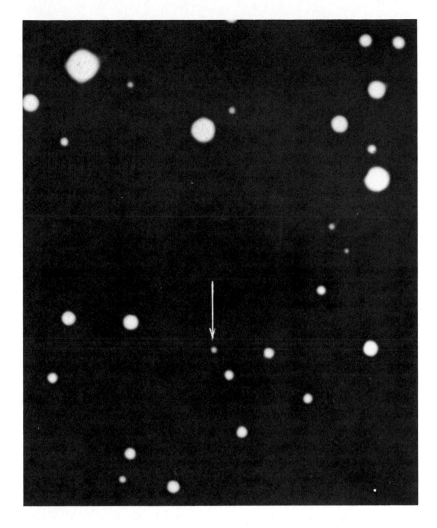

Figure 9-5   The Quasar OH471
This quasar has one of the biggest redshifts ever discovered. This redshift corresponds to a speed slightly over 90 percent of the speed of light. According to the Hubble law, OH471 must be 16 billion light years away. (Copyright by the National Geographic Society—Palomar Sky Survey. Reproduced by permission from the Hale Observatories.)

tronomers realized that they possessed photographs of quasars dating back many years. In fact, these objects had been ignored as ordinary stars for decades. Some old photographs of quasars date back to the 1880s.

By carefully examining the images of quasars on old photographic plates, astronomers soon realized that quasars rapidly fluctuate in brightness. A typical quasar might flare up, increasing its brightness by 50 to 100 percent in only a week. Week-to-week variations in brightness are quite common.

Recall that one of the important precepts of all modern science is that nothing can travel faster than light. In Chapter 7 we saw that this principle can be used to place strict limits on the maximum possible size of an object in space. Specifically, an object cannot vary its brightness more rapidly than the light-travel time across that object. Because quasars vary their brightness from week to week, a typical quasar cannot be larger than a light week across (that's about 200 billion kilometers, or roughly fifteen times the diameter of Pluto's orbit).

To summarize the dilemma of quasars, we first realize that quasars (or at least the energy-producing regions inside quasars) are very small. This fact is inferred from their rapid brightness variations. We also realize that quasars are extremely luminous. This fact is inferred from their huge redshifts, which means that they are located at vast distances from the earth. These two facts conspire to make quasars the most extraordinary and perplexing objects ever discovered in the sky. Consequently, astrophysicists are confronted with the staggering task of trying to explain how the energy output of a hundred galaxies can be generated in a volume only a few times as large as our solar system.

When quasars were first discovered, there seemed to be an unbreachable chasm between ordinary galaxies and these mysterious, compact, superluminous objects. But during recent years, astronomers have found that many unusual galaxies have properties that bridge the gap. For example, look at the galaxy called M82 in Figure 9-6. Filaments of gas are easily seen extending tens of thousands of light years outward from the galaxy's nucleus. In addition, the galaxy is a strong source of radio waves, and its nucleus is emitting copious

Figure 9-6 The Peculiar Galaxy M82 (also called NGC 3034 or 3C231)
Huge filaments of gas erupting out of this galaxy are evidence of a colossal explosion that occurred roughly 2 million years ago at the galaxy's center. This galaxy is also a powerful source of radio waves, X rays, and infrared radiation. (Hale Observatories.)

X rays. Surely there are some extraordinary energetic processes occurring at the galaxy's nucleus.

The irregular galaxy M82 is in the northern sky, in the constellation of Ursa Major. Its counterpart in the southern sky is the peculiar galaxy called NGC 5128 in the constellation of Centaurus. As you can see in Figure 9-7, there is a vast, fragmented dust lane stretching across the face of the galaxy. The shattered appearance of this immense dust lane bears witness to some sort of violent events that must have occurred (are still occurring?) at the galaxy's nucleus. Indeed, the nucleus of NGC 5128 is a strong source of X rays. In addition, vast quantities of radio waves are coming from two huge regions of the sky located above and below the galaxy along a line perpendicular to the dramatic dust lane. Indeed, NGC 5128 is one of the brightest sources of radio waves in the entire sky.

Objects like M82 and NGC 5128 clearly demonstrate that there can be more to galaxies than a rotating collection of widely spaced stars. Something quite unusual must be occurring at the centers of each of these galaxies. But this is child's play compared with the kinds of galaxies that apparently bridge the gap to quasars. These interlopers are called *Seyfert galaxies,* named after the astronomer Carl Seyfert, who first drew attention to some of their unusual properties in the early 1940s.

A fine example of a Seyfert galaxy is shown in Figure 9-8. This galaxy, called NGC 1275, is located in the constellation of Perseus and is a strong source of both X rays and radio waves. In fact, NGC 1275 outshines all other X-ray and radio sources in that part of the sky. And look at its extraordinary shape! Vast filaments of gas stretch outward in all directions for thousands upon thousands of light years. Indeed, this disrupted galaxy reminds us of the Crab Nebula (recall Figure 3-5), clearly suggesting that violent explosive processes are occurring at the galaxy's core.

An examination of the cores of Seyfert galaxies demonstrates why they may be a "missing link" between ordinary galaxies and quasars. Seyfert galaxies, such as M77 in Figure 9-9 or NGC 4151 in Figure 9-10, all have extremely bright nuclei. Their cores are excessively luminous compared with the surrounding spiral arms. In fact, the dazzling nuclei of these galaxies was one of the key features that

Figure 9-7   The Peculiar Galaxy NGC 5128
*This exotic galaxy is a powerful source of radio waves and
X rays. It is located about 10 million light years away, in the
direction of the constellation of Centaurus. The galaxy's distorted
appearance suggests that an explosive event or process may have
occurred at the galaxy's center. (Hale Observatories.)*

165

**Figure 9-8   The Exploding Galaxy NGC 1275
(also called 3C84)**
This remarkable photograph shows filaments of gas erupting from
the center of a very distorted galaxy. This galaxy is also a powerful
source of X rays and radio waves. (Courtesy of Roger Lynds, Kitt
Peak National Observatory.)

166

**Plate 9   The Andromeda Galaxy (also called M31
or NGC 224) in Andromeda**
*A typical galaxy measures 100,000 light years across and
contains several hundred billion stars. This nearby spiral galaxy is
the largest member of the Local Group. It is located 2¼ million
light years from the earth and is just barely bright enough to be seen
with the naked eye. Two small companion galaxies also appear in
this wide-angle view. (Copyright by the California Institute of
Technology and the Carnegie Institution of Washington.
Reproduced by permission from the Hale Observatories.)*

Plate 10   The Large Magellanic Cloud
*This nearby, irregularly shaped galaxy is only 160,000 light years
from the earth. It is the nearest member of the Local Group and is a
companion galaxy to our own Milky Way Galaxy. It can easily be
seen with the naked eye from southern latitudes. The galaxy
contains a gigantic emission nebula, called 30 Doradus, seen near
the left side of the photograph. (Copyright by the Association of
Universities for Research in Astronomy, Inc. The Cerro Tololo
Inter-American Observatory.)*

***Plate 11   The Whirlpool Galaxy (also called M51 or NGC 5194)
in Canes Venatici***
*The arching spiral arms of a galaxy are outlined by huge glowing
clouds of gas called* emission nebulas. *This dramatic spiral galaxy
is roughly 15 million light years from the earth. A companion galaxy
(called NGC 5195) is located at the end of a luminous bridge
connecting the two galaxies. This bridge of glowing gases was
created during a near-collision between the two galaxies some 50
million years ago. (Courtesy of the U.S. Naval Observatory.)*

Plate 12   The Spiral Galaxy NGC 7217 in Pegasus
*Galaxies are seen in every unobscured part of the sky. Roughly
three-quarters of the brightest galaxies in the sky are spiral galaxies,
like the one shown here. Our own Milky Way Galaxy is a spiral
galaxy, with the sun located about two-thirds of the way from the
center to the edge. If we could view our Galaxy from a great
distance, it would probably look very similar to the galaxy shown in
this photograph. (Courtesy of the U.S. Naval Observatory.)*

Plate 13   The Spiral Galaxy NGC 253 in Sculptor
*Galaxies are the largest individual objects in the universe.
Unfortunately, astronomers still do not fully understand how and
why galaxies formed. For some reason, the material of the early
universe coalesced into huge lumps called* protogalaxies *that
contain just enough matter to become a galaxy. Trying to understand
the birth of a protogalaxy is one of the challenging topics in
modern theoretical astrophysics. (Copyright by the California
Institute of Technology and the Carnegie Institution of
Washington. Reproduced by permission from the Hale
Observatories.)*

Plate 14 The Spiral Galaxy NGC 7331 in Pegasus
Galaxies are rotating. All the stars and nebulas in a galaxy revolve about the galaxy's center. It often takes over a hundred million years for a star to complete one orbit of its galaxy's nucleus. For example, it takes the sun 250 million years to travel once around the center of our Galaxy. (Copyright by the California Institute of Technology and the Carnegie Institution of Washington. Reproduced by permission from the Hale Observatories.)

Plate 15   The Peculiar Galaxy M82 (also called NGC 3034) in Ursa Major

Billowing clouds of gas and dust extend all across this disrupted galaxy. Powerful X rays, radio waves, and copious infrared radiation are pouring out of the galaxy's nucleus. Huge filaments of gas (most easily seen in Figure 9-6) are erupting out of the galaxy's center. All this activity is evidence of a recent explosion that is tearing the galaxy apart. This strange galaxy is one member of a nearby cluster of galaxies located only 10 million light years from the earth. (Copyright by the California Institute of Technology and the Carnegie Institution of Washington. Reproduced by permission from the Hale Observatories.)

Plate 16   The Exploding Galaxy NGC 5128 in Centaurus
*A dramatic dust lane cuts across the face of this exploding galaxy.
Vast quantities of radio waves are pouring out of extended regions
to either side of the dust lane. X rays are coming from the galaxy's
center. Cosmic violence of this magnitude can be most easily
explained by the presence of a supermassive black hole at the
galaxy's nucleus. This galaxy is only 13 million light years from
the earth. (Copyright by the Association of Universities for Research
in Astronomy, Inc. The Cerro Tololo Inter-American Observatory.)*

Figure 9-9   The Seyfert Galaxy M77 (also called NGC 1068)
*This galaxy has an unusually bright, almost starlike, nucleus. It is
an extremely powerful source of infrared radiation. Infrared
radiation from the galaxy's nucleus varies from week to week. (Lick
Observatory.)*

167

Figure 9-10   The Seyfert Galaxy NGC 4151
*The nucleus of this galaxy is so bright that if it were very far away,
it would probably be mistaken for a quasar. There is strong
evidence for the violent ejection of gas from the nucleus of this
galaxy. (Hale Observatories.)*

Carl Seyfert pointed out in 1943. More significantly, the brilliant star-like nuclei of Seyfert galaxies often exhibit variability over periods of months or weeks. All the energy of these starlike nuclei is therefore being produced in a very small volume.

Suppose that these Seyfert galaxies, such as NGC 1068 or NGC 4151, were much farther away from us. Instead of 100 or 200 million light years, suppose they were located ten times as far away, at distances of a few billion light years. Surely we would not be able to see any spiral arms. Photographs with our most powerful telescopes would reveal only the bright starlike nuclei. Obviously these galaxies would be mistaken for quasars.

There are no nearby quasars. All quasars have substantial redshifts, which means that they are far away. This is unfortunate, because we cannot examine them at close range. Indeed, at close range we find Seyfert galaxies, exploding galaxies, and active galaxies, all characterized by extraordinary processes occurring at their cores. Perhaps by carefully examining some of these nearby objects we can gain important insights into their remote and elusive quasar cousins.

The nearest "rich" cluster of galaxies is located in the constellation of Virgo. This vast cluster is only 50 million light years away and covers 120 square degrees of the sky, as seen from the earth (that is about 160 times as large as the area covered by the full moon). Over 1,000 galaxies are easily identified within this portion of the sky, and certainly many thousands of dwarf ellipticals have completely escaped detection.

As is the case with most rich clusters, the central region of the Virgo cluster is dominated by an enormous elliptical galaxy called M87. An exceptionally fine photograph of this colossus appears in Figure 9-11.

M87 has several interesting characteristics. It is a powerful source of radio waves. A strong source of X rays is located at its center. And it has a distinctly starlike nucleus. Even more intriguing is a large luminous "jet" surging up from the starlike nucleus. This jet is most easily seen on short time exposure photographs, such as those in Figure 9-12, which also clearly reveal the galaxy's starlike nucleus. The entire situation is clearly reminiscent of the jet associated with the quasar 3C273 shown in Figure 9-4.

169

**Figure 9-11   The Supergiant Elliptical Galaxy M87
(also called NGC 4486 or 3C274)**
This enormous galaxy in the Virgo cluster is a powerful source of
radio waves and X rays. This long time exposure reveals hundreds
of faint star clusters that surround the galaxy like a swarm of bees.
The distance to this extraordinary galaxy is 60 million light years.
(Kitt Peak National Observatory.)

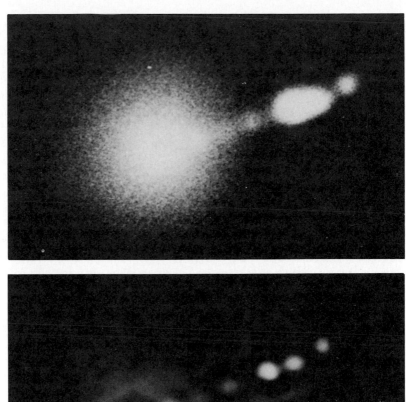

Figure 9-12   The "Jet" in M87
A short time exposure reveals a luminous jet surging out of the nucleus of M87. The upper view is a single, ordinary photograph of the jet. The lower view is a computer-enhanced exposure made from several photographic plates. (Courtesy of H. C. Arp and J. J. Lorre, Hale Observatories.)

During 1977, two teams of astronomers performed a series of detailed observations of M87. One team carefully measured the brightness across the galaxy. Of course, the dazzling starlike nucleus was prominent. In trying to explain their data, the astronomers concluded that there must be a strong, compact source of gravity at the galaxy's center. They argued that this unseen yet extremely powerful source of gravity causes the stars to crowd together around the galaxy's center. This stellar congestion is the reason we see a starlike galactic nucleus in photographs such as Figure 9-12. According to the astronomers' calculations, the mass of this invisible source of gravity must be 5 billion solar masses.

The second team of astronomers focused their attention on spectroscopic observations. They carefully examined the shapes of spectral lines across the galaxy. Looking toward the galaxy's center, they found that the spectral lines are unusually broad. This means that stars near the galaxy's nucleus are moving with unusually high speeds. In trying to explain their data, this second team of astronomers also concluded that a strong, compact source of gravity must be located at the galaxy's center. All the slow-moving stars have long ago been swallowed by this unseen object. Only the swiftly moving stars orbiting this powerful source of gravity have managed to survive. Once again, calculations revealed that the mass of this invisible source of gravity must be 5 billion solar masses.

The obvious implication is that a *supermassive black hole* exists at the core of M87. This was not the first time that astronomers contemplated the possibility of extremely massive black holes. For many years it has been realized that supermassive black holes have many attractive features for explaining quasars and exploding galaxies. First of all, even supermassive black holes are quite small. For example, a 1-billion-solar mass black hole has a diameter of only 5½ light hours. In addition, an enormous amount of energy must be tied up in the hole's vast gravitational field. The small size plus the high energy content meet all the necessary requirements for explaining quasars. But how can this energy get out? What kind of mechanisms might remarkably occur to tap the hole's vast supply of gravitational energy?

Unfortunately, we are far from any final answers. Nevertheless, some important and promising theoretical breakthroughs have recently occurred as a result of the work of R. V. E. Lovelace at Cornell and R. D. Blandford at Cambridge. Their ideas involve a disk of gases about a supermassive black hole (quite similar to the disk in the Cygnus X-1 system) *plus* the galaxy's magnetic field.

Imagine a supermassive black hole, like the one that might exist at the center of an active galaxy or a quasar. As we argued in Chapter 5, we would naturally expect the hole to be sucking up interstellar gas and dust in its vicinity. Because of the hole's rotation, this infalling material is preferentially confined to the hole's equatorial plane (that is, perpendicular to the hole's axis of rotation). The result is a huge disk, like a gigantic version of the disk that surrounds the black hole in Cygnus X-1.

All galaxies have magnetic fields. These magnetic fields that thread the space between the stars are extremely weak because they are spread out and diluted across millions upon millions of cubic light years. But the gases spiraling in toward a supermassive black hole certainly carry along some of this galactic magnetic field. Over the ages, the magnetic field in the disk can become extremely concentrated. A cross-sectional diagram showing the arrangement of the hole, the disk of gases, and the magnetic field is given in Figure 9-13.

The phenomena of electricity and magnetism are intimately related. Consequently, as the concentrated magnetic field at the inner edge of the disk whips around the hole, a powerful electric field is produced. This field is produced in much the same way that the whirling armature of an electric generator creates an electric current in a power station here on the earth. The cosmic analog is therefore properly called a *black hole dynamo.*

This relativistic dynamo surrounds the black hole with an enormous electric field. Nowhere else in the universe does nature see electric fields of this colossal magnitude and intensity. Calculations reveal that this electric field contains so much energy that, according to the equation $E = mc^2$, this energy is transformed into vast quantities of matter and antimatter. A perpetual deluge of electrons and antielectrons spews outward from the hole's electrified environ-

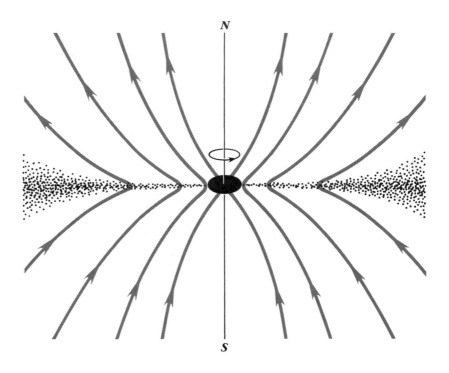

Figure 9-13 A Magnetized Accretion Disk Surrounding
a Supermassive Black Hole
*Gases captured by a rotating black hole form a disk in the plane of
the hole's equator. These gases also carry some of the galaxy's
overall magnetic field. As the gases spiral inward toward the hole,
the magnetic field becomes extremely concentrated and is aligned
parallel to the axis of rotation of the hole.*

ment. Although many details remain to be worked out, it seems clear that this outpouring of particles and energy could account for the brilliance of quasars and the violent activity at the cores of exploding galaxies.

Quasars and active galaxies have puzzled astronomers since their discovery in the early 1960s. Finally, the enigma is beginning to yield to concerted theoretical and observational investigations. Both avenues of research are strongly pointing to the remarkable possibility that supermassive black holes provide the enormous energy requirements of quasars and exploding galaxies. Of course, it would be ironic and paradoxical that the most luminous objects in the universe are powered by huge black holes from which nothing can escape.

a

Figure 9-14   The Galactic Center in Sagittarius
*The center of our Galaxy (indicated by the white rectangle) is
located 30,000 light years away, in the direction of the
constellation of Sagittarius. In 1979, two Harvard astronomers
announced their discovery of rapidly moving gases at the galactic
center, which suggests the presence of a supermassive black hole
(estimated mass: 5 million solar masses). In addition, astronomers
operating a gamma-ray-detecting satellite announced their
discovery of copious gamma rays (energy: 511 keV) coming from
the galactic center. This is exactly the energy we should expect from
the annihilation of electrons and antielectrons that would be
spewed out from a black hole dynamo. Is there a supermassive
black hole at the center of our Galaxy? Is it possible that
supermassive black holes lie hidden at the centers of many
ordinary-looking galaxies across the universe? (Hale
Observatories.)*

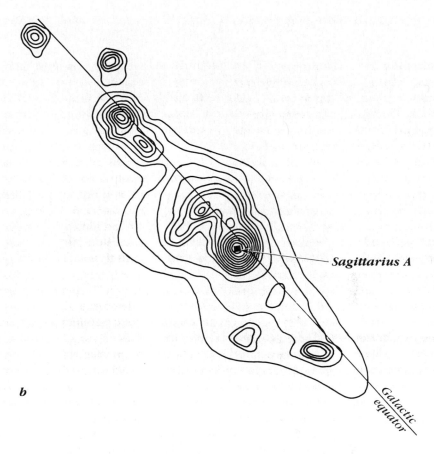

b

Sagittarius A

Galactic
equator

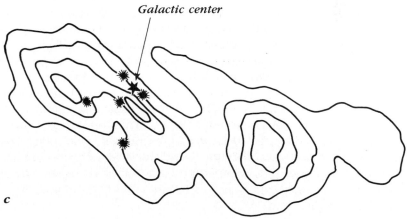

Galactic center

c

177

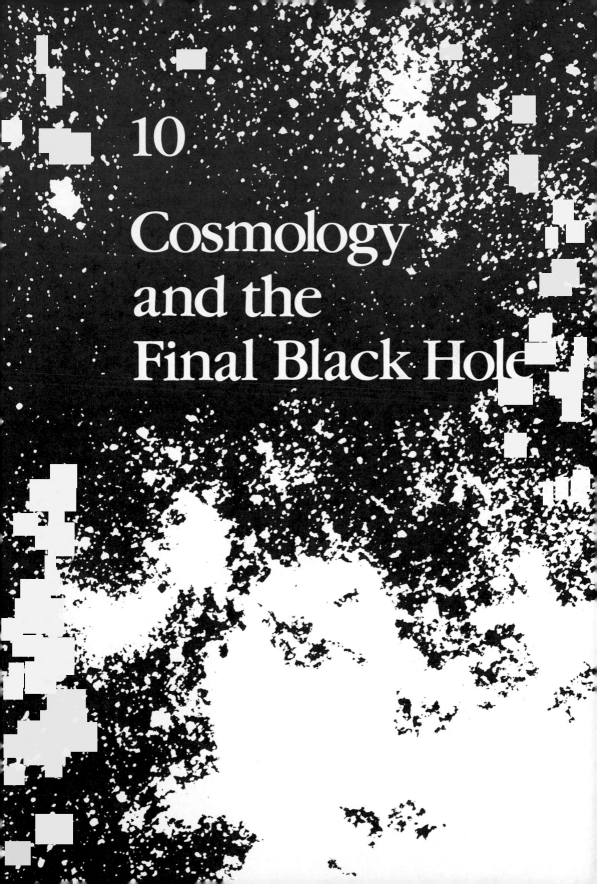

10

Cosmology
and the
Final Black Hole

Silently, effortlessly, vast clusters of galaxies glide across space, coasting farther and farther apart. No force propels these colossi in their blind and futile plunge toward the nonexistent edge of the cosmos. All their impetus was imparted in one inconceivable blow dealt during the creation of the universe itself.

We live in an expanding universe. The distances between widely separated clusters are increasing. And the rate at which these clusters are moving apart is proportional to their separation. This is what "expansion" means. This is the simplest and most straightforward interpretation of the Hubble law.

But surely this cosmic expansion must be slowing down. All those galaxies scattered across space are exerting gravitational forces on each other. The mutual gravitational attraction among all the galaxies must be decelerating the frantic pace of universal expansion.

Will the expansion of the universe ever stop? Or will the rate of deceleration be too small to effectively inhibit the headlong rush of the cosmos perpetually toward the infinitely distant reaches of space? To answer these questions, to discover the ultimate fate of the universe, we must first construct theoretical models of the universe. Because gravity is the force that affects the evolution of the cosmos, we must be sure to use our best theory of gravity in our computations. Our models will therefore be based on the general theory of relativity. And, finally, we must examine motions of the most distant galaxies. We must try to accurately measure the rate at which the expansion of the universe is decelerating. By matching this deceleration with one of our relativistic models, we can calculate the destiny of the cosmos. In this way, the stars reveal our most distant future.

From the viewpoint of general relativity, you never need to speak about the "force" of gravity. Instead, Einstein's theory explains that gravity curves space and time. The stronger the gravity, the greater the curvature of spacetime. Objects such as planets and light rays simply move along the shortest paths in curved spacetime. This is the true meaning of general relativity: matter tells spacetime how to curve, and curved spacetime tells matter how to behave.

There is a lot of matter in the universe. Obviously, all this matter must have an effect on the geometry of spacetime. The uni-

Figure 10-1   Galaxies and Geometry
*Vast quantities of matter are spread across space. All matter carries
gravity, and all gravity warps the geometry of space and time. The
matter of the stars and galaxies should therefore give the universe
an overall shape. A deviation from perfect flatness would, however,
be extremely slight. (Kitt Peak National Observatory.)*

verse as a whole must have a *shape* because of all the matter in the universe. And furthermore, this geometry must influence the behavior of the matter across the cosmos. Consequently, the shape of the universe must be intimately related to the ultimate future of the universe. A universe that expands forever must have a very different shape from a universe whose expansion eventually stops and whose contraction begins. But what do we really mean by the "shape" of the universe?

Imagine that you shine two powerful laser beams out into space. Now suppose that you align these two beams so that they start off perfectly parallel. Finally suppose that nothing gets in the way of these two beams; we can follow them for billions of light years across the universe, across the space whose curvature we wish to detect.

There are only three possibilities. First, we might find that our two beams of light remain perfectly parallel, even after traversing billions of light years. In this case, space is not curved. The universe has *zero curvature* and space is *flat*.

Alternatively, we might find that our two beams of light gradually converge. The two beams gradually get closer and closer together as they move across the universe. Indeed, the two beams might eventually intersect at some enormous distance from the earth. In this case, space is not flat. Just as lines of longitude on the earth's surface intersect at the poles, the geometry of the universe must be like the geometry of a sphere. We therefore say that the universe has *positive curvature* and space is *spherical*.

The third and final possibility is that the two parallel beams of light eventually diverge. The two beams gradually get farther and farther apart as they move across the universe. In this case, the universe must also be curved. But it must be curved in the opposite sense of the spherical case. We therefore say that the universe has *negative curvature*. In the same way that a sphere is a positively curved surface, a saddle is a good example of a negatively curved surface. Just as parallel lines drawn on a sphere always converge, parallel lines drawn on a saddle always diverge. Mathematicians say that saddle-shaped surfaces are "hyperbolic." Thus, in a negatively curved universe, we say that space is *hyperbolic*.

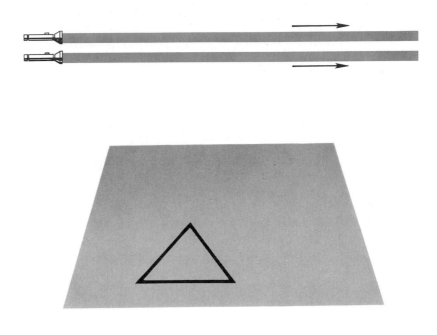

Figure 10-2   Flat Space
If two parallel light beams remain parallel forever, then space has zero curvature. The geometry of flat space is the three-dimensional analogy of a two-dimensional plane. Geometrical properties of flat space include the fact that the sum of the angles of a triangle is exactly equal to 180 degrees.

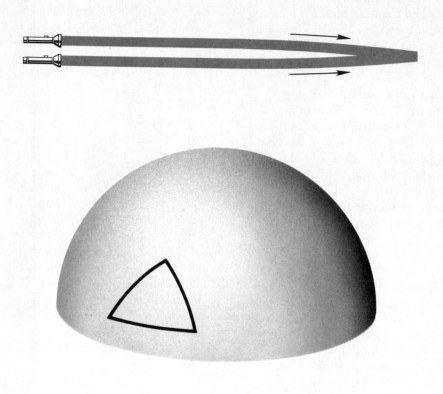

Figure 10-3   Spherical Space
If two parallel light beams eventually converge, then space has positive curvature. The geometry of positively curved space is the three-dimensional analogy of the two-dimensional surface of a sphere. Geometrical properties of spherical space include the fact that the sum of the angles of a triangle is always greater than 180 degrees.

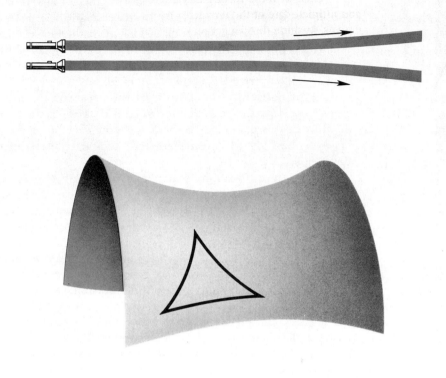

Figure 10-4   Hyperbolic Space
*If two parallel light beams eventually diverge, then space has
negative curvature. The geometry of negatively curved space is the
three-dimensional analogy of the two-dimensional surface of a
saddle. Geometrical properties of hyperbolic space include the fact
that the sum of the angles of a triangle is always less than 180
degrees.*

Each of these three cases corresponds to a different behavior and ultimate fate of the universe. To understand these different alternatives, imagine throwing a rock up into the air. Once again there are only three possibilities. First, the rock may simply go up and come back down. In that case, the speed of the rock was *less* than the escape velocity from the earth.

Alternatively, you could throw the rock upward with a much greater speed, perhaps with the aid of a rocketship. If the speed of the rock *equals* the escape velocity from the earth, then the rock will never fall back. It will just manage to escape from the earth's gravitational pull.

The third possibility is that the rock is given a speed *greater* than the escape velocity from the earth. In this case, the rock has no trouble escaping the earth's gravity. Even after traveling an enormous distance, we find that the rock continues its journey away from the earth at a considerable speed.

By analogy with the rock, our question about the future of the universe becomes, "Are clusters of galaxies rushing apart with speeds great enough to overcome their mutual gravitational attraction?"

Cosmological models of the universe based on Einstein's general theory of relativity were first worked out by the Russian mathematician Alexandre Friedmann in 1922. Friedmann found that if there is enough matter in the universe to stop the expansion, then there is enough gravity to cause space to fold around on itself like a sphere. Thus, in a positively curved universe, the expansion eventually stops and contraction begins. In a spherical universe, the speeds of the galaxies are less than their mutual escape velocity.

If the galaxies are rushing apart with speeds that exactly equal their mutual escape velocity, then the universe will never collapse back upon itself. Of course, the expansion of the universe eventually slows to a snail's pace. But there is not quite enough gravity to cause the galaxies to stop completely in their tracks. Consequently, according to Friedmann's models, there is not quite enough gravity to cause space to fold around on itself. This universe, therefore, has zero curvature. In a flat universe, the galaxies just barely manage to overcome their mutual gravitational attraction.

Finally, perhaps the mutual gravitational attraction between clusters of galaxies is so weak that the universal expansion will continue with vigor infinitely far into the future. According to general relativity, the shape of this universe must be hyperbolic. In a hyperbolic universe, clusters of galaxies will be rushing apart forever and ever.

In discussing the shape and fate of the universe, we are dealing with properties of the universe on an enormous scale. Individual galaxies are lost in the background as we gaze across distances far greater than the separations between superclusters. It is like looking at the palm of your hand. You know that you actually are made of an enormous number of tiny atoms, each of which consists of a dense nucleus orbited by electrons in otherwise empty space. But all this detail is lost as you run your fingers over your skin. In the same fashion, the universe is remarkably smooth and featureless in the enormous scale we are considering. In the scale of the Friedmann models, the universe is homogeneous and isotropic.

Because of the homogeneity of the universe on the largest scale, we can speak intelligently about the *average density* of the universe. We can act as though matter is uniformly spread across space, because individual details of one galaxy or another are totally insignificant. The advantage is that, by speaking about the average density, we can relate the amount of matter across space to the gravity that curves space. But we have just seen how the shape of the universe is intimately associated with the fate of the universe. Consequently, a measurement of the average density across the cosmos should enable us to predict the ultimate future.

The flat (zero curvature) model universe constitutes the dividing line between the spherical (positively curved) universe and the hyperbolic (negatively curved) universe. Consequently, the average density associated with a flat universe is the crucial quantity that differentiates the three cases. The average density of a flat universe is therefore called the *critical density.*

The critical density is simply the average density our universe should have *if* it is really flat. With this density, the cluster of galaxies will just barely manage to overcome their mutual gravitational attrac-

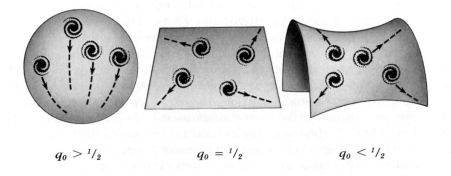

$$q_0 > {}^1/_2 \qquad q_0 = {}^1/_2 \qquad q_0 < {}^1/_2$$

Figure 10-5   Geometry and Destiny
The ultimate fate of the universe is directly related to the geometry of the universe. If space is spherical, universal expansion will someday stop and contraction will begin. If space is flat, the universe will just barely manage to expand forever. If space is hyperbolic, the expansion will continue forever with vigor.

tion. If there is a little more matter spread across space than the critical density, then there will be enough gravity to someday stop the expansion of the universe. If there is a little less matter spread across space than the critical density, then the universe will have no trouble expanding forever and ever.

Obviously, the critical density must be related directly to the rate at which the universe is expanding. After all, the critical density is the quantity that decides whether or not the expansion will continue forever. We measure the expansion rate of the universe simply by observing redshifts and distances of remote galaxies. The same data that were used to plot the Hubble law in Figure 8-13 give an expansion rate corresponding to a *critical density of three hydrogen atoms per 1,000 liters of space.* (That is the same as 6 kilograms of matter per billion billion cubic kilometers.)

The relationships between the various quantities that affect the future of the universe are summarized in the table on the following page. A spherical universe is said to be "closed" in the same sense that the surface of a sphere closes around upon itself. In principle, if you traveled for a long time in a particular direction, you would eventually get back to your starting place, just like an adventurous sailor who circumnavigates the globe. Such a universe is finite. It does not extend forever and ever. Nevertheless, it does not have an edge or a boundary. Neither does it have a center. After all, you could travel forever around the earth and never get to the "edge" or the "center." In the same way, an astronaut could journey forever across a closed, spherical universe and never reach the "edge" or the "center."

In contrast to the spherical case, both flat and hyperbolic universes are infinite. They extend forever in all directions and are said to be "open." Actually, we might say that a flat universe is "just barely open," because this case is the dividing line between a closed, spherical universe and a wide-open, hyperbolic universe. Of course, these open universes do not have any edges or centers simply because they extend forever. Consequently, no matter which universe we live in, it is always meaningless to ask, "What is beyond the edge of the universe?" And because none of these universes has any edges or boundaries, it is also meaningless to ask, "What is the universe expanding into?" Such questions are fundamentally nonsensical.

Geometry of space	Curvature of space	Average density throughout space	Deceleration parameter	Type of universe	Ultimate future of the universe
Spherical	Positive	Greater than the critical density	Greater than ½	Closed	Eventual collapse
Flat	Zero	Exactly equal to the critical density	Exactly equal to ½	Flat	Perpetual expansion (just barely)
Hyperbolic	Negative	Less than the critical density	Between 0 and ½	Open	Perpetual expansion

In trying to decide which universe we live in, we find that these relativistic Friedmann cosmologies really depend on only two important numbers. First of all, the cosmological models must depend on the rate at which the universe is expanding. The number that expresses this expansion rate is called the *Hubble constant* and is usually given the symbol H_0. Allan Sandage at Hale Observatories has spent many years carefully measuring the redshifts and distances of remote galaxies in an exhaustive effort to determine this expansion rate. His value for the Hubble constant is 17 kilometers per second per million light years. This simply means that, as you gaze out into space, you pick up 17 kilometers per second of universal expansion for each million light years. For example, a galaxy that is 100 million light years away should be receding from us at a speed of 1,700 kilometers per second.

The second important number expresses the rate at which the universal expansion is slowing down. This number is called the *de-*

celeration parameter and is usually given the symbol q_0. For a totally empty, wide-open, hyperbolic universe, the deceleration parameter equals zero ($q_0 = 0$). This is the extreme case. Quite simply, there is no matter or gravity to slow down the expansion at all.

For a flat universe, where there is just enough matter to ensure that the clusters of galaxies just manage to escape from each other, the deceleration parameter equals one-half ($q_0 = \frac{1}{2}$). And for closed, spherical universes, the deceleration parameter is greater than one-half. The greater the deceleration parameter, the sooner expansion stops and collapse begins.

The history of the universe for various deceleration parameters is given schematically in Figure 10-6. Because it would probably be confusing and ambiguous to talk about the "size" of the universe, we instead speak of the *scale of the universe.* The scale of the universe is simply any very large distance across space, such as the distance between two widely separated clusters of galaxies. Figure 10-6 shows how this very large distance changes with time, depending on the size of the deceleration parameter. If the deceleration parameter equals zero, the universe expands forever with absolutely no deceleration at all. In this case, the age of the universe must be 20 billion years. This age is deduced simply by extrapolating backward through time, assuming that the expansion rate has not slowed down over the years.

Actually, the universe must be slightly younger than 20 billion years old. The universe is not empty. The gravitational interactions between clusters of galaxies must be producing some deceleration. Consequently, the deceleration parameter must be larger than zero, and, as indicated in Figure 10-6, the Big Bang must have occurred less than 20 billion years ago. For example, if the universe is flat ($q_0 = \frac{1}{2}$), then the age of the universe should be 13 billion years. If the universe is closed, then the Big Bang occurred less than 13 billion years ago. If the universe is open, its age is between 13 and 20 billion years.

There are several ways of estimating the size of the deceleration parameter, or an equivalent quantity such as the average density or curvature. Unfortunately, none of these methods is very reliable, and conflicting conclusions are often obtained.

191

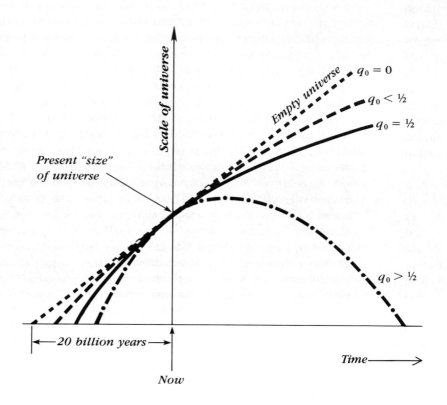

Figure 10-6   The History of the Universe
*This graph shows how the scale of the universe changes with time
for various values of the deceleration parameter (q_0).*

One method involves trying to estimate the average density throughout space. By observing the motions of galaxies, we can deduce how much matter they contain. We then calculate what the average density would be if all this matter were smoothly spread across the universe. The answer is much less than the critical density. Consequently, the unverse should be open and hyperbolic. The expansion should continue forever with vigor.

A second method involves trying to guess the age of the universe. This can be done by searching for the oldest stars (such as those in globular clusters) and trying to determine their ages. Alternatively, it is possible to estimate the age of the universe by measuring the relative abundances of certain ancient radioactive isotopes. From both of these approaches, we find that the age of the universe is between 10 and 18 billion years. This is disappointingly inaccurate. It does not allow us to decide between an open or a closed cosmology.

A third method involves measuring the amount of deuterium left over from the creation of the cosmos. Deuterium is an isotope of hydrogen. It is often called "heavy hydrogen" and constitutes an important stepping-stone in the chain of thermonuclear reactions that lead from hydrogen to helium. Most astronomers feel quite certain that a substantial fraction of the primordial hydrogen was converted into helium soon after the Big Bang. During this process, some deuterium was left over; it got left behind in the chain of events that led from light hydrogen to helium when the cosmic thermonuclear reactions turned off. The amount of this primordial deuterium is very sensitive to the density and rate of expansion during the early universe. Measurements of the abundance of deuterium from satellite observations strongly suggest that the average density is much less than the critical density. Once again, this means that we live in an open, hyperbolic universe.

Yet another method involves trying to measure the deceleration of the universe directly. This is done by measuring the redshifts and distances to a large number of distant galaxies. These data are then plotted on a Hubble diagram, such as Figure 8-13. If there has been a large amount of deceleration, then the universe must have been expanding faster in the past than it is today. If there has been a small amount of deceleration, then the universe must have been expanding at a rate comparable to today's rate. These differences show up by extending the Hubble diagram to include data from the most remote galaxies, as shown in Figure 10-7.

This method should give a direct determination of the deceleration parameter. Unfortunately, like all the other methods, there is considerable uncertainty in the observations, as evidenced by the scatter of the data points on the graph in Figure 10-7.

The most courageous recent attempt to measure the deceleration parameter by this method was made by Jerome Kristian, Allan Sandage, and James Westphal at Hale Observatories. Their data, published in 1978, suggest a value of $q_0 = 1.6$. This is far above the critical value for a flat universe ($q_0 = \frac{1}{2}$) and suggests that we live in a closed, positively curved universe. Indeed, the present age of the universe should be only 12 billion years, and the expansion of the universe will stop in about 60 billion years. This would be followed by a collapse of the universe and another Big Bang roughly 130 billion years from now.

If the universe is closed, then we are living inside a black hole. Indeed, the entire universe itself is one inconceivably vast black hole that encompasses all space and time. The history of the universe then simply consists of erupting from a *past singularity* 20 billion years ago, only to plunge into a *future singularity* 130 billion years from now. It is like traveling straight up the middle of the Penrose diagram that we examined in Figure 6-5.

If there is enough matter across space to stop the expansion of the universe, then this same density of matter also suffices to envelop the entire universe inside its own cosmic black hole. Redshifts will someday turn into blueshifts as our expanding universe gradually evolves into a collapsing universe. Remote clusters of galaxies — once so widely separated — will crowd in on each other. Densities and pressures within this shrinking cosmos get higher and higher until everything is crushed out of existence at the cosmic singularity extending across space at the end of time.

The cosmic black hole in which the universe swallows itself is fundamentally different from any of the black holes we have examined so far. With ordinary black holes or supermassive black

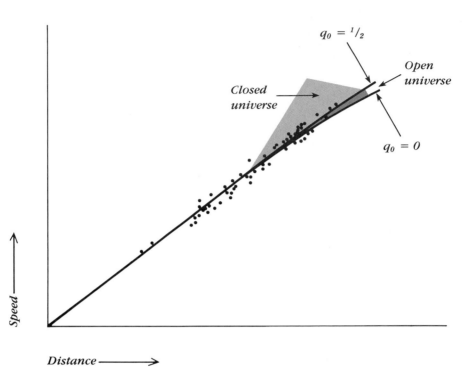

Figure 10-7   Hubble Curves and Deceleration Parameters
*It should be possible to deduce the size of the deceleration
parameter from measurements of the redshifts and distances of the
most remote galaxies we can find. If the data fall between the
curves marked $q_0 = \frac{1}{2}$ and $q_0 = 0$, then we live in an open
universe that will expand forever. If the data fall above the curve
marked $q_0 = \frac{1}{2}$, then we live in a closed universe. Unfortunately,
the data do not clearly favor either case. (Adapted from
A. Sandage.)*

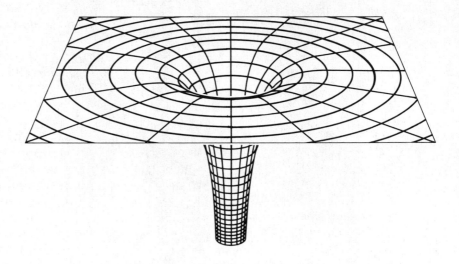

**Figure 10-8   Asymptotically Flat Spacetime
Surrounding a Black Hole**
*An ordinary black hole is smoothly connected to the flat spacetime
of the rest of the universe. The existence of this flat spacetime far
from the hole allows us to speak intelligently about the hole's mass,
charge, and spin.*

196

holes, there is always plenty of flat spacetime far from the hole. Although the hole itself is a region of extreme spacetime curvature, this warped volume is always smoothly connected to the flat spacetime far from the hole, as diagramed in Figure 10-8. Theoretical physicists therefore speak of *asymptotically flat spacetime* far from any ordinary black holes.

As we discussed in Chapter 5, black holes devour material in an unforgiving, irreversible fashion. Many of the properties and qualities of infalling matter are permanently removed from the universe. Nevertheless, we saw that the three basic quantities are preserved: mass, charge, and spin. These quantities survive because they can be measured by people standing far from the hole. These quantities are well defined in the flat spacetime that surrounds the hole. Indeed, the existence of asymptotically flat spacetime is the primary reason why we can talk intelligently about the mass, charge, and spin of a black hole in the first place.

If the universe is closed, then we are doomed to plunge down the cosmic black hole that envelops all space and time. As the collapsing universe swallows itself, there is obviously *no* asymptotically flat spacetime onto which the cosmic black hole is connected. With no distant frame of reference, even the most fundamental properties of matter are completely destroyed as the universe rushes headlong to a state of infinite pressure, infinite density, and (most significantly!) infinite spacetime curvature.

Nothing can survive passage through infinitely warped spacetime. The most basic quantities in science become forever lost. The most fundamental numbers in physics that describe the most intimate details of matter and radiation become indeterminable. This is why we cannot intelligently ask what happened *before* the Big Bang roughly 20 billion years ago. Such regions of spacetime are completely cut off from us by the infinitely warped spacetime of the past singularity from which our universe was born. Even the basic structure of matter in the "previous universe" could have been vastly different from the atoms of which our present universe is composed. And the "next universe" — the universe that will be born from the condemned ashes of our cosmos — will be totally unrecognizable.

11
Primordial
Black Holes

Where did the earth and the heavens come from? How did it all begin? When was the universe created? And why was it created? These are some of the most basic and profound questions ever formulated by the human mind. They directly reflect the natural curiosity and inquisitiveness that characterize the human being. This inquisitiveness, this desire to understand the world around us, is one of the crucial attributes that dramatically differentiate us from all other forms of life on our planet.

In ancient times, many of these fundamental questions were answered with myth and fantasy. In the absence of any significant data, our ancestors populated the heavens with gods and demons. Most cosmological questions, especially those involving the creation of the universe, were explained as the direct, divine action of one or more superhuman beings.

Fifty years ago, humanity was treated to two important astronomical discoveries that have direct bearing on these age-old cosmic issues. In the mid-1920s, we discovered that galaxies are scattered across the universe. These enormous stellar whirlpools are seen in every unobscured part of the sky. And five years later, significant data had accumulated about the overall motions of these galaxies. By 1930, it was clear that widely spaced galaxies are rushing farther and farther apart. This means that we live in an expanding universe. It also means that the universe must have been born from an infinitely dense state that existed some 15 to 20 billion years ago. This cosmic birth must have been initiated by a violent primordial explosion, called the Big Bang, that occurred throughout all space at the beginning of time. Indeed, the Big Bang was an explosion of space itself. This is the most straightforward and obvious interpretation of all our data. And even today we see clear evidence of this explosion as we watch the fragments—widely separated clusters of galaxies—plunging headlong into the blackened void of cosmic space.

In spite of these important astronomical discoveries, it is sometimes astonishing how much we have yet to learn. For example, we are not at all sure how and why galaxies formed in the first place. We understand how a cloud of interstellar gas and dust fragments and condenses into stars. We also have a good understanding of planet

200

formation: how bits of rock and ice can accompany the birth of a star, leaving behind the debris that coalesces into asteroids, planets, and their satellites. But no one has succeeded in developing an equivalently complete and detailed theory of galaxy formation. Galaxies are the largest individual objects in the universe. A typical galaxy measures 100,000 light years across and contains an amount of matter equal to several hundred billion suns. For some unknown reason, the early universe was destined to fragment into enormous clumps that eventually became galaxies.

As we noted in the previous chapter, the universe is remarkably smoothed out and uniform on the largest scales. If we gaze across hundreds of millions of light years, across distances so vast that individual galaxies are too tiny to be noticed, we find that the universe is extremely homogeneous in all directions. This means that the very early universe must also have been very smooth and uniform in all directions. But it could *not* have been *perfectly* smooth. If the universe had been perfectly homogeneous right after the Big Bang, then it would be perfectly smooth today. There would be no galaxies, no stars, neither planets nor people. So there must have been some lumpiness, some slight degree of inhomogeneity across the face of the newborn cosmos.

Regardless of any detailed theory of galaxy formation, it is clear that galaxies must have originated from some slight lumpiness that existed throughout the universe right after the Big Bang. Of course, if a primordial lump is to become a galaxy, it must contain or gravitationally attract an amount of matter equal to the mass of a galaxy (that is, about 200 or 300 billion solar masses). This may sound as though we need some enormous lumps to make galaxies, but this is not so. The newborn universe was so incredibly dense and compact that only slight deviations from perfect smoothness sufficed to produce lumps containing the required quantities of matter. For example, calculations show that a galaxy-sized enhancement of only 1 percent above the average density of the universe should have been sufficient to create a galaxy.

In any reasonable theory of density fluctuations, it is always easier to make little lumps than big lumps. Consequently, for every galaxy-sized lump that existed right after the Big Bang, there must

Figure 11-1   A Cluster of Galaxies
*Clusters of galaxies are scattered far and wide across the universe.
Although the universe is remarkably smooth and homogeneous on
the very largest scale, the existence of galaxies means that there was
some lumpiness in the universe immediately after the Big Bang.
(Kitt Peak National Observatory.)*

have been vast numbers of smaller lumps, each containing small amounts of matter. In addition, the density enhancement inside these tiny lumps could have been much more pronounced than in the larger lumps. Whereas a galaxy-sized lump would involve only a 1 percent enhancement over the surrounding density, a 1-billion-ton lump (roughly the mass of an asteroid) might contain a 100-percent enhancement. The large-scale (1 percent) density fluctuations in the newborn universe eventually evolved into galaxies containing all the stars, nebulas, planets, and satellites we see in the sky. But what happened to all those tiny density fluctuations, especially to all those lumps that contained substantial amounts of matter (100-percent enhancements) compared with their surroundings?

The early universe must have been extremely hot and extremely dense. We conclude this simply by extrapolating back toward the moment of the Big Bang. For example, at one hundredth of a second after the Big Bang, the temperature must have been 100 billion degrees and the average density roughly 4 billion grams per cubic centimeter (the same as 70,000 tons per cubic inch) throughout all space.

In 1971, Stephen W. Hawking at the University of Cambridge suggested that, under the crushing conditions of the early universe, all those tiny over-dense lumps could be compressed to form numerous tiny black holes! A similar suggestion had been made a few years earlier by the Soviet physicists Ya. B. Zel'dovich and I. D. Novikov, but it was Hawking's analysis that inspired scientists to consider seriously the possibility that the universe is populated with vast numbers of very tiny *primordial black holes.*

This is a truly astounding prediction. Prior to Hawking's work, everyone firmly believed that black holes could be created only by the deaths of massive stars. Since low-mass stars (like the sun) become white dwarfs and moderate-mass stars become neutron stars at the end of their lives, only the most massive stellar corpses possess enough matter to overpower their internal pressures and completely collapse to become black holes. As we learned at the end of Chapter 3, the minimum mass of one of these "classical black holes" is 2½ solar masses. But we now realize that prominent, small-sized density fluctuations in the newborn universe could have been crushed to become tiny black holes. As the early universe vigorously expands, the small-sized, over-dense lumps are left behind with enough matter to effectively inhibit participation in the universal expansion. Under appropriate conditions of pressure and density, these lumps can then implode to form primordial black holes. This scenario was confirmed with detailed computer calculations in 1979 by two teams of Soviet astrophysicists, including Zel'dovich and Novikov. It is possible that black holes with masses as low as a hundred-thousandth of a gram could have been created in this fashion.

Unlike their huge stellar cousins, these tiny primordial black holes possess some amazing properties. To understand these properties, we must first appreciate how very small these primordial black holes really are. For example, imagine a primordial black hole weighing a billion tons. That is roughly the mass of a typical mountain here on the earth. But in the case of a billion-ton primordial black hole, this amount of matter has been compressed inside its event horizon, which has a diameter of one ten-trillionth of a centimeter (10^{-13}cm). That is roughly the size of a proton. We therefore see that typical primordial black holes contain the masses of mountains and are roughly the size of subatomic particles.

In dealing with the submicroscopic world, we must make use of a branch of physics called *quantum mechanics.* Quantum mechanics describes how particles such as electrons and protons behave. It is the branch of physics that is used to calculate the structure of atoms and the interactions of nuclei. Since primordial black holes have subatomic dimensions, we must be sure to account for quantum effects.

There are significant differences between the ordinary world around us and the submicroscopic world of quantum mechanics. In the ordinary, macroscopic world we have no trouble knowing where things are. You know where your house is; you know where your car is; you know where your mother is. But when you peer into the subatomic world of electrons and nuclei, you can no longer speak with this same confidence and surety. A certain amount of fuzziness and uncertainty enter into the description of reality. Because of the incredibly small dimensions in the quantum world, things are no longer crystal clear.

To appreciate the reasons for this fuzziness or uncertainty, imagine trying to measure the position of a single electron. In order to find out where the electron is located, you have to see it. In order to see it, you shine a light on it. But since the electron is so tiny and has such a small mass, the photons in your beam of light possess enough energy to give the electron a mighty kick. As soon as a photon strikes the electron (so that you can see it to measure its position), this same photon imparts a huge momentum to the electron, causing it to recoil in some direction. Consequently, in trying to measure the location of an electron with great precision, you necessarily introduce a large amount of uncertainty into the speed or momentum of that electron.

These ideas are at the heart of the famous *Heisenberg uncertainty principle,* first formulated by Werner Heisenberg in 1927. Quite simply, there is a reciprocal fuzziness between position and momentum. The more precisely you try to measure the position of a particle, the more unsure you are of how the particle is moving. And conversely, if you try to accurately determine the speed of a particle, you find that you are no longer quite sure where the particle is. This same fuzziness does not enter into our ordinary, everyday world simply because macroscopic objects (such as your house, car, or mother) have enormous masses compared with electrons and are therefore not easily kicked around by the sunlight that shines on them.

There is also an analogous uncertainty or fuzziness involving energy and time. You cannot know the exact energy of a quantum mechanical system with infinite precision at every moment in time. Over short time intervals, there can be large uncertainty in the amounts of energy in the subatomic world.

Obviously we can look upon the Heisenberg uncertainty principle as merely an unfortunate limitation in our ability to know about the nature of reality. But it is considerably more than that. Instead of simply bemoaning our fate of not being able to know everything with infinite precision, we can instead explore how the uncertainty principle might unlock new doors in our understanding of the universe.

One of the important conclusions of Einstein's special theory of relativity is that mass and energy are equivalent. This equivalence is expressed by the famous equation $E = mc^2$. But one version of the uncertainty principle explains that there is a reciprocal fuzziness between energy and time. In view of $E = mc^2$, this principle can be restated as a reciprocal fuzziness between mass and time in the quantum world. In other words, in a very brief interval of time, we cannot be quite sure how much matter there is in a particular location — even in *empty* space. During the brief moment that nature blinks, particles and antiparticles can spontaneously appear and disappear.

There is nothing mysterious about antimatter. A particle and an antiparticle are identical in almost every respect, except that they carry opposite electric charges. For example, an antielectron has a positive charge, whereas an ordinary electron carries a negative charge. By always creating particles and antiparticles in equal numbers, Nature ensures that electric charge is not magically produced out of nothingness. As long as particles and antiparticles spontaneously appear or disappear in pairs, Nature's electric charge balance sheet is always satisfied.

The time intervals over which this process occurs are incredibly brief. For example, during an interval of one billion-trillionth of a second (that is, 10^{-21} second), an electron and an antielectron can spontaneously appear and disappear. And this can occur absolutely *anywhere.* It can also occur with more massive particles, but because of the "reciprocal" nature of the uncertainty principle, these more massive particles can exist only for correspondingly shorter time intervals. For example, the proton is 2,000 times as heavy as the electron. Therefore pairs of protons and antiprotons can appear and disappear provided they last for only 1/2,000 as long as the pairs of electrons and antielectrons.

One of the functional doctrines of subatomic physics (usually attributed to Murray Gell-Mann of Caltech) is, "If it is not strictly forbidden, then it must occur." "It" refers to any quantum process. Therefore, pairs of every conceivable particle and antiparticle are constantly being created and destroyed everywhere, at every location all across the universe. Of course we have no way of directly observing these pairs of particles and antiparticles. That is forbidden by the uncertainty principle; the particle-antiparticle pairs simply exist for such short time intervals that direct observation is impossible. For this reason, they are called *virtual pairs.* They don't "really" exist; they "virtually" exist.

Although virtual pairs of particles and antiparticles cannot be directly observed, their effects have been detected. Imagine an electron in orbit about the nucleus of an atom, such as a hydrogen atom.

Ideally, the electron should follow its orbit in a smooth and unhampered fashion. But because of the constant brief appearance and disappearance of pairs of particles and antiparticles, tiny electric fields exist for extremely short intervals of time. These tiny, fleeting electric fields cause the electron to jiggle slightly in its orbit. This jiggling was first detected in 1947 by W. E. Lamb and R. C. Retherford, who noticed a tiny shift in the spectrum of light from the hydrogen atom. This phenomenon is called the *Lamb shift*. It provides powerful support for the idea that every point in space—all across the universe—is seething with virtual pairs of particles and antiparticles.

The concept of virtual pairs also provides a ready explanation for the phenomenon of *pair production*. For many years it has been known that highly energetic gamma rays can convert their energy into pairs of particles and antiparticles according to the equation $E = mc^2$. Quite simply, the gamma ray disappears (upon colliding with a second photon), and in its place a particle and an antiparticle appear. This process is commonly observed in high-energy nuclear experiments. Indeed, it is one way in which nuclear physicists can manufacture many exotic species of particles and antiparticles.

Where do these particles and antiparticles come from? They spring from Nature's ample supply of virtual pairs. The gamma ray provides a virtual pair with so much energy that the virtual particles can materialize and appear as *real* particles in the real world. The only requirement is that you satisfy Nature's balance sheet. If you want to create a particle and an antiparticle of total mass m, then your incoming gamma ray must possess an amount of energy E that at least satisfies the condition $E = mc^2$. If your photons carry too little energy, the reaction will not proceed. Conversely, the more energetic the photon, the more massive are the particles and antiparticles that can be manufactured.

Until 1974, no one ever believed that any of these quantum mechanical processes could possibly have any effect on black holes. Once again Stephen Hawking turned the tables.

Imagine a tiny primordial black hole, one that contains a billion tons of completely collapsed matter. As we learned, this black hole is roughly the same size as a proton. And we also know that virtual pairs of particles and antiparticles are constantly being created and destroyed all around the black hole. But because of the limitations of the uncertainty principle, we cannot be absolutely sure where any of these quantum processes is occurring. For example, consider a virtual pair that appears right alongside the tiny black hole. For a fleeting moment, a particle and antiparticle separate. Because of the hole's small size, one of these two particles can find itself *inside* the hole. Obviously the hole eats this particle. Its partner is abandoned. Its partner has been deprived of its alter ego and therefore cannot quantum mechanically vanish behind the fuzzy veil of the uncertainty principle. The abandoned particle is therefore forced to become a real particle in the real world! Since this particle is outside the black hole, it is free to wander off to distant portions of the universe. As seen by a remote observer watching this process, a particle has come from the black hole. We are therefore led to the absolutely astounding realization that *black holes emit particles!*

Hawking then proceeded to follow this line of reasoning to its logical conclusion by pointing out that Nature's energy balance sheet *must* be satisfied. The energy to create these particles must come from somewhere. The obvious source is the energy of the hole's gravitational field. As the black hole emits particles, it must lose energy, and therefore (since $E = mc^2$) its mass must decrease. Quite simply, for every kilogram of particles emitted, the mass of the black hole must decrease by exactly 1 kilogram. Consequently, *black holes evaporate!*

In pursuing these ideas with mathematical calculations, Hawking showed that the magnitude and extent of these effects depend critically on the mass of the black hole. The smaller the hole, the more pronounced are the quantum effects. To see why this is so, imagine an ordinary big black hole, such as the collapsed corpse of a massive burned-out star. The diameter of the event horizon is several miles, and the hole's powerful gravitational field extends over a large volume of space. Of course, all this space is seething with virtual pairs. Fleeting pairs of particles and antiparticles are constantly separating and recombining. Because of the extreme brevity of this phenomenon, a virtual particle and its antiparticle never get very far apart. Although we cannot know exactly where they are, we feel quite certain that their separation is small. Because of this small separation, it is highly improbable that, during the brief moment of separation, one member of a virtual pair would find itself outside the hole's gravitational domain. Consequently, the probability of particles escaping from an ordinary black hole is exceedingly small.

In contrast, this phenomenon is more pronounced for smaller black holes that could only have been created in the moments immediately following the Big Bang. As we learned in Chapter 5 (recall Figure 5-2), the diameter of the event horizon is directly related to the mass of the hole. The smaller the mass, the smaller the diameter of the event horizon and, consequently, the smaller the extent of the hole's gravitational domain.

Consider a virtual pair that momentarily appears inside the event horizon of one of these small primordial black holes. Because of the hole's small size, there is now a real possibility that one member of the pair could find itself outside the hole during the brief moment of separation. This particle would be free to escape from the hole, thereby robbing the hole of some of its matter. If the primordial hole is very tiny, this phenomenon becomes very pronounced. For example, if the hole is roughly the size of a proton, then a virtual pair that momentarily appears inside the event horizon has a high probability that one of its members finds itself outside the hole. One of these tiny black holes would therefore radiate particles (or, equivalently, energy) at a furious rate.

The rate at which a black hole radiates particles (or energy) can be described by assigning a temperature to the hole. Big, ordinary black holes must have extremely low temperatures, because it is highly improbable that particles could quantum mechanically escape from their extensive gravitational fields. Indeed, the temperature of a 1-solar-mass black hole is a millionth of a degree above absolute zero.

Smaller black holes have higher probabilities of emitting particles and therefore have correspondingly higher temperatures. A primordial hole containing a trillion tons of matter has a temperature of 1 billion degrees. The smaller the hole, the higher its temperature. This relationship between a hole's mass and its temperature is given in Figure 11-2.

Think about a tiny black hole that is emitting particles and energy at a particular temperature. As material flows out of the hole,

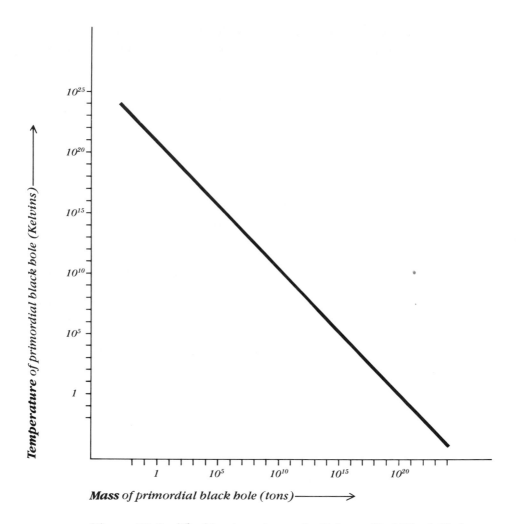

Figure 11-2   The Temperature of a Primordial Black Hole
*A black hole radiates particles and energy corresponding to a
temperature that depends only on the hole's mass. The smaller the
hole, the higher its temperature. For example, a 1-billion-ton black
hole radiates at a temperature of roughly 1 trillion degrees. This
graph displays this inverse relationship between the mass of a hole
and its temperature.*

the mass of the hole decreases. Therefore its temperature goes up, and therefore it radiates even faster, thereby decreasing its mass even more.

Obviously this is a catastrophic, runaway process. The smaller it gets, the hotter it gets. The hotter it gets, the more it emits. The more it emits, the smaller it gets. And so on and so on in a vicious cycle that dooms the black hole to completely evaporate and *explode!* During the final tenth of a second, the black hole gives up the remainder of its mass in one incredible burst of gamma rays equivalent to the simultaneous detonation of 10 million 1-megaton hydrogen bombs! This amount of energy would be sufficient to totally pulverize the moon and scatter the powdered lunar fragments far and wide across the solar system.

This process of black hole explosions means that there are no tiny black holes around today. Tiny primordial holes are so hot and are therefore evaporating so furiously that they cannot long survive. For example, a 1-million-ton primordial black hole completely evaporates and explodes after only 30 years. Thirty years after the Big Bang, all primordial holes with mass less than a million tons have exploded in violent bursts of gamma rays. Bigger black holes are cooler and are therefore radiating particles and energy at a slower rate. They survive for longer times. For example, a 1-billion-ton primordial black hole lasts for nearly 300 million years. The relationship between the masses and lifetimes of black holes is given in Figure 11-3.

Since the universe is roughly 20 billion years old, we can use this relationship between mass and lifetime to predict the masses of primordial holes that could have survived to the present time. The answer is 4 billion tons. Primordial black holes with masses greater than 4 billion tons are sufficiently cool and are therefore evaporating sufficiently slowly that they should still be around today.

How many of these small black holes might we expect to find floating across space? It's almost anybody's guess. Of course, if conditions of pressure and density immediately following the Big Bang were not right, then none of these black holes formed and there would be none around today. But on the other hand, from everything we know about the universe, we cannot rule out the possibility that these black holes are extremely plentiful. Some reasonable estimates

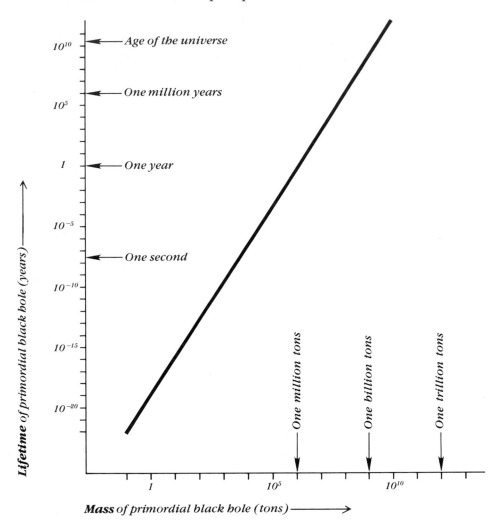

Figure 11-3   The Lifetime of a Primordial Black Hole
As a black hole radiates, it loses matter. As its mass decreases, its temperature goes up, and therefore the hole radiates still more furiously. This vicious cycle causes the hole to completely evaporate and explode. This graph shows how the lifetime of a black hole depends on its mass. Black holes with masses greater than 4 billion tons should be cool enough to have survived to the present time.

are so high that there could be one or two of these primordial holes right here in our own solar system! Unknown, unseen, these powerful compact sources of energy orbit the sun along with all the other planets, asteroids, and meteoroids.

This possibility is sufficiently reasonable and intriguing that serious proposals have been made to fly a gamma-ray telescope onboard the space shuttle in the 1980s. This telescope would be designed to search the skies for high-energy (10 MeV) gamma rays that should be coming from an evaporating black hole weighing a few billion tons.

The implications of finding one of these holes in our solar system are staggering. In all probability, it would be technologically feasible to bring the hole back to the earth. After all, the hole would have a mass roughly comparable to the mass of a small asteroid, and for many years there has been a realistic discussion about mining the asteroid belt. The primordial hole could be placed into orbit about the earth. Its gamma rays could be converted into microwaves that are beamed to receiving stations on the ground, where the radiation is further transformed into electrical power. Hawking estimates that the power output of just one hole would be 6,000 megawatts, equivalent to six large nuclear power plants.

One of these earth-orbiting primordial black holes would also make thermonuclear weaponry totally obsolete. Long before any lumbering intercontinental ballistic missiles could get out of their silos, the powerful beam of microwaves could be focused on military installations with the speed of light. Under the full intensity of the 6,000 megawatts, the ground would be transformed into a seething mass of boiling lava.

A less intense beaming of the microwaves from the black hole could be used for population control. Teeming masses of humanity could be exposed to a constant, low-level dosage. These microwaves would induce sufficient chromosome breakage and genetic defects that production of live offspring would be rendered impossible.

In any of these scenarios, it would be extremely important to be sure that the mass of the primordial hole is greater than 4 billion tons. With a lower mass, the hole would be in danger of completely evaporating and exploding. Such an event would have a devastating impact on the environment. Indeed, it has been proposed that a nearby black hole explosion millions of years ago produced a sufficient amount of radiation to cause the extinction of the dinosaurs.

Of course, there is always the possibility that conditions immediately following the Big Bang did *not* favor the creation of primordial black holes. In that case, the smallest black holes are formed from the corpses of burned-out stars and contain at least 2½ masses of completely collapsed matter. As we saw, these holes are still extremely cold (typical temperatures are a millionth of a degree above absolute zero), and no black hole evaporations have yet occurred anywhere in the universe. In fact, because the cosmic microwave background is 3 degrees above absolute zero, all these ordinary, big black holes have been absorbing more radiation than they are emitting.

But someday all this will change. For example, by extending Figure 11-3 to very large (stellar) masses, we find that a 10-solar-mass black hole (like the one in the Cygnus X-1 system) has a lifetime of 10 million trillion trillion trillion trillion trillion years (10^{67} years). In this far-off age, stellar black holes will be exploding in violent bursts of gamma rays. Of course this would occur only if we live in an "open" universe that continues to expand forever. Only then would the universe survive long enough so that large black holes would be in danger of completely evaporating. Long before then, other significant changes would have occurred. In only 10 trillion years from now, all the hydrogen and helium — the gases from which newborn stars are made — will have been completely used up. From that time on, no new stars will be formed. And all the old stars will have become cold white dwarfs, neutron stars, or black holes, depending on their masses. Then the universe will be dark — oppressively dark — for billions upon billions of years, until that far-off age when widely scattered black hole stellar corpses briefly light up the sky one final time.

There is a problem with extrapolating our ideas this far into the future. To appreciate this problem, simply recall that black holes eat stuff. In doing so, the black hole removes information from the universe. For example, all the data about the chemical composition, the color, the shape and texture of infalling material is completely destroyed upon plunging into a black hole. Indeed, as we saw in

Chapter 5, only the mass, charge, and spin of the infalling matter are preserved. For this reason, black holes are aptly termed "information sinks." They are places where information about infalling material forever leaves the universe.

With Hawking's discovery of black hole evaporation, we realize that particles can indeed get out of black holes. Of course this quantum mechanical process is important only for tiny black holes. Nevertheless, this outpouring of matter means that primordial black holes look like *white holes*. A white hole is simply a black hole running backward in time. It is a place from which things erupt (rather than a place into which things fall). Indeed, in 1975, Stephen Hawking succeeded in proving that at the quantum level, primordial black holes are indistinguishable from white holes!

As material pours out of a primordial hole, new information is being introduced into the universe. In principle, the matter erupting from one of these holes carries color, texture, chemical composition—all fresh, new data that never before existed in the universe. A primordial (black/white) hole is therefore an "information source."

The quantum processes that underlie Hawking's evaporation mechanism are fundamentally random. Because of the uncertainty principle, we cannot predict where and when an individual particle might appear. Therefore the data that are injected into the universe from one of these holes are also fundamentally random. This is the essence of Hawking's recently formulated *randomicity principle*. Like Heisenberg's uncertainty principle, it is also a statement of basic limitations on our ability to learn about reality. If there are evaporating primordial holes scattered across space, then new particles, new data, and new information are pouring into the universe in a totally random, unpredictable fashion. As scientists try to understand the game of the universe, the cards and the players come and go with hardly any attention to any rules.

Albert Einstein never liked quantum mechanics, even though he played an important role in the foundation of the subject. The unpredictability and uncertainty of quantum physics was repugnant to him. He expressed his feelings of revulsion by saying, "God does not play dice."

Although this book is a testament to his genius, on this one point Einstein was wrong. Quantum mechanics works. There is an inherent uncertainty in the quantum world. But in view of Hawking's discoveries, there may be a level of randomicity that extends even across the universe. As Stephen Hawking put it: "God not only plays dice but also sometimes throws them where they cannot be seen."

Index